Smart Sensors, Measurement and Instrumentation

Volume 31

Series editor

Subhas Chandra Mukhopadhyay
Department of Engineering, Faculty of Science and Engineering
Macquarie University
Sydney, NSW
Australia
e-mail: subhas.mukhopadhyay@mq.edu.au

More information about this series at http://www.springer.com/series/10617

Robert LeMoyne • Timothy Mastroianni
Donald Whiting • Nestor Tomycz

Wearable and Wireless Systems for Healthcare II

Movement Disorder Evaluation and Deep Brain Stimulation Systems

 Springer

Robert LeMoyne
Department of Biological Sciences and
Center for Bioengineering Innovation
Northern Arizona University
Flagstaff, AZ, USA

Donald Whiting
Department of Neurosurgery Allegheny
General Hospital
Allegheny Health Network Neuroscience
Institute
Pittsburgh, PA, USA

Timothy Mastroianni
Independent
Pittsburgh, PA, USA

Nestor Tomycz
Department of Neurosurgery Allegheny
General Hospital
Allegheny Health Network Neuroscience
Institute
Pittsburgh, PA, USA

ISSN 2194-8402   ISSN 2194-8410   (electronic)
Smart Sensors, Measurement and Instrumentation
ISBN 978-981-13-5807-4   ISBN 978-981-13-5808-1   (eBook)
https://doi.org/10.1007/978-981-13-5808-1

Library of Congress Control Number: 2017952009

This Springer imprint is published by the registered company Springer Nature Singapore Pte Ltd.
The registered company address is: 152 Beach Road, #21-01/04 Gateway East, Singapore 189721,
Singapore

To my Wife, thank you for everything. Love Always.
I would like to thank my Mother, Father, and Brother for their support.
"Nothing transcends the power of the human spirit."
from a homeless Vietnam Veteran and very loyal friend
And in the wind, he's still alive.
"To beat a tiger, one needs a brother's help."
Chinese Proverb
Thanks Tim.
Chiri mo tsumoreba yama to naru.
A favorite Japanese Proverb (in Romaji)
Translation:
Even dust piled up becomes a mountain.

Preface

The domain of wearable and wireless systems for biomedical and movement disorder treatment applications, such as through smartphones and portable media devices, is anticipated to grow exponentially. Even from the time of presenting the draft manuscript of this book to the time of publication, the prevalence of this subject is expected to undergo meaningful transformation and evolution. These devices enable wireless inertial sensor applications to an assortment of scenarios pertaining to movement disorder evaluation and deep brain stimulation systems.

Since 2010, when LeMoyne, Mastroianni, and our research team applied a novel smartphone application for quantifying Parkinson's disease tremor and gait quantification in the context of a wireless accelerometer, the opportunities have expanded considerably. The authors are delighted to provide a contribution to this exciting field with the anticipation of the considerable array of developments in years to follow. Please enjoy the knowledge and intellectual inspiration that our book provides with the goal of providing meaningful, robust, and optimal rehabilitation for many.

Flagstaff, AZ, USA
Pittsburgh, PA, USA

Robert LeMoyne
Timothy Mastroianni
Donald Whiting
Nestor Tomycz

Contents

Chapter 1
Wearable and Wireless Systems for Movement Disorder Evaluation and Deep Brain Stimulation Systems

Abstract The implementation of wearable and wireless systems for deep brain stimulation offers the opportunity to substantially advance the treatment of progressive neurodegenerative movement disorders, such as Parkinson's disease and Essential tremor. Deep brain stimulation offers an efficacious alternative regarding scenarios for which the intervention by medication has become intractable while avoiding the permanency of ablative neurosurgery. Even subsequent to the expert surgical application of the deep brain stimulation system, the acquisition of the optimal parameter configuration is inherently resource intensive and challenging in nature. With the advent of wearable and wireless systems, the response to therapy intervention for movement disorder status can be objectively quantified through the inertial sensor signal, such as an accelerometer and gyroscope. Furthermore, wireless connectivity to the Internet enables experimental and post-processing resources to be remotely situated effectively anywhere in the world. With machine learning amended to the post-processing capability, clinical diagnostic acuity is substantially advanced. Foundational subjects are elucidated, such a general perspective regarding Parkinson's disease and Essential tremor, traditional ordinal methodologies for diagnosing severity, and the development of deep brain stimulation including surgical techniques for installation. The role of wearable and wireless systems, such as the smartphone, for quantifying the status of neurodegenerative movement disorders, such as Parkinson's disease and Essential tremor, is presented. The utility of applying machine learning for augmenting diagnostic acuity of movement disorder status is addressed. The integration of wearable and wireless systems, such as the smartphone, with machine learning is discussed for the ability to distinctively classify between deep brain stimulation set to "On" and "Off" status for Parkinson's disease and Essential tremor. The amalgamation of wearable and wireless systems with deep brain stimulation using machine learning as an augmented post-processing application implicate the evolutionary trends for the ability to achieve closed-loop optimization of parameter configurations with the development of Network Centric Therapy for a quantum leap in the treatment intervention for neurodegenerative movement disorders, such as Parkinson's disease and Essential tremor.

© Springer Nature Singapore Pte Ltd. 2019
R. LeMoyne et al., *Wearable and Wireless Systems for Healthcare II*,
Smart Sensors, Measurement and Instrumentation 31,
https://doi.org/10.1007/978-981-13-5808-1_1

Keywords Wearable and wireless system · Smartphone · Wireless inertial sensor · Accelerometer · Gyroscope · Deep brain stimulation system · Movement disorders · Parkinson's disease · Essential tremor · Parameter configuration · Optimization · Closed-loop tuning · Network Centric Therapy

1.1   Introduction

The efficacious treatment of movement disorder is a matter of rampant concern. In particular, Parkinson's disease and Essential tremor are two progressive neurodegenerative movement disorders that have been diagnosed throughout the span of modern medicine [1–3]. These movement disorders impact the lives of millions of people [4, 5]. Furthermore, their incidence is more prevalent for older people [2, 5, 6].

For many decades Parkinson's disease and Essential tremor were treated by medication [2, 4, 7–9]. Upon medication strategies becoming intractable neurosurgery targeting the deep brain is an alternative generally of last resort [2, 4, 10]. In recent decades deep brain stimulation provides a unique alternative therapy strategy [11, 12].

Deep brain stimulation offers a unique therapy for the robust treatment of an assortment of movement disorders, such as Parkinson's disease and Essential tremor. A vast assortment of parameters is provided to the clinical specialist for the determination of an appropriate configuration, such as the quantity of active electrodes, frequency of simulation, the stimulation amplitude, and the pulse width [11, 12]. Even after the surgical implantation of a deep brain stimulation system by a skilled neurosurgeon, the capability to ascertain an optimal parameter configuration presents a challenge, which is imperative for the successful application of a deep brain stimulation system [13]. A logical algorithmic approach for converging upon a parameter configuration is desirable [14].

Another concern is the traditional means of evaluating the severity of movement disorders, such as Parkinson's disease and Essential tremor. Traditional means of assessment involve the expert although subjective interpretation of movement disorder status [15–21]. An issue with these ordinal scale techniques is the inability to produce consistently congruent reliability studies, which furthermore have been shown to demonstrate variations in reliability among disparate levels of expertise for members of the medical community [22–25]. LeMoyne et al. advocate the alternative strategy of applying wearable and wireless sensors for the objective and quantified measure of movement disorder characteristics, such as tremor [26–32].

The objective of the book is to provide a perspective leading to the amalgamation of wearable and wireless systems and deep brain stimulation for the optimization of patient-specific therapy. These developments represent the development of a quantum leap in medical treatment for movement disorders, such as Parkinson's disease and Essential tremor, devised by LeMoyne et al. known as Network Centric Therapy. Network Centric Therapy respective of movement disorder represents the confluence of the Internet of Things through wearable and wireless systems and deep brain stimulation for the movement disorder. In essence, Network Centric Therapy evolves the treatment of movement disorders, such as Parkinson's disease and Essential tremor, to the domain of data science.

1.2   Perspectives of the Chapters

A brief perspective of the pending chapters of the book is provided. The reader should be aware of the authors' observation that the subject matter of each individual chapter could literally be expanded to the magnitude of a book. Furthermore, the chapters can be segmented into multiple respective sections.

1.2.1   Perspective of Chap. 2: Movement Disorders: Parkinson's Disease and Essential Tremor—A General Perspective

The preliminary chapter pertains to a brief summary of the relevant neurodegenerative movement disorders under consideration throughout the span of the pending chapters: Parkinson's disease and Essential tremor. Casual observation of Parkinson's disease and Essential tremor may infer that they are similar. However, they are fundamentally different with regard to the neurological basis of the disorder, traditional diagnosis, and medical therapy approach [1, 2, 4, 7–10, 16, 18–21].

Parkinson's disease is predominantly attributed to dysfunction of structures of the basal ganglia, which is attributed to degeneration of dopaminergic neurons regarding the substantia nigra [4, 33]. The severity of Parkinson's disease is clinically determined through an ordinal rating scale, such as the Unified Parkinson's Disease Rating Scale (UPDRS) and evolved MDS-UPDRS (Movement Disorder Society-Unified Parkinson's Disease Rating Scale) [16, 18, 19]. Parkinson's disease is conventionally treated with medical intervention, such as Levodopa [1, 8, 9]. Regarding scenarios for which medical therapies become intractable, both Parkinson's disease and Essential tremor can be treated by lesioning neurological structures of the deep brain [1, 2, 10].

By contrast the neurological foundation for Essential tremor is not conclusively resolved [2, 7]. The Fahn-Tolosa-Marin Tremor Rating Scale is a standard ordinal scale approach for the clinical assessment of Essential tremor [20]. Furthermore, a common medical intervention for Essential tremor is through the prescription of Propranolol [2, 7].

Recent developments have advocated the breakthrough technology of deep brain stimulation for the treatment of Parkinson's disease and Essential tremor, but the clinician is presented with a considerable assortment of parameter configurations in the quest to ascertain the optimal settings [12, 13, 34]. LeMoyne et al. have researched, developed, tested, and evaluated the use a wearable and wireless inertial sensor systems, such as a smartphone, for assessing the efficacy of deep brain stimulation [35–37]. The amalgamation of deep brain stimulation for treating movement disorder with wearable and wireless inertial sensor systems presents the opportunity for the development of Network Centric Therapy.

1.2.2 Perspective of Chap. 3: Traditional Ordinal Strategies for Establishing the Severity and Status of Movement Disorders, Such as Parkinson's Disease and Essential Tremor

Clinical determination of movement disorder severity is traditionally determined through the expert although subjective evaluation according to an ordinal scale [26]. Clinicians generally apply the Unified Parkinson's Disease Rating Scale (UPDRS) and evolved MDS-UPDRS (Movement Disorder Society-Unified Parkinson's Disease Rating Scale) for the five-point ordinal quantification of severity for Parkinson's disease [16, 18, 19]. Respective of Essential tremor, the Fahn-Tolosa-Marin Tremor Rating Scale represents a five-point ordinal approach to quantify the severity of Essential tremor [20, 21]. Multiple other ordinal scale strategies exist for assessing movement disorder severity. However, the difficulty to translate between these rating methodologies is a matter of concern [15]. Furthermore, the consistent reliability of these ordinal rating techniques is a topic of contention [15, 24, 25].

An objectively quantified alternative for evaluating movement disorder has been considerably researched, developed, tested, and evaluated by LeMoyne et al. for an assortment of movement disorders, such as hemiplegic spasticity, Parkinson's disease, and Essential tremor. This technique applies wearable and wireless inertial sensors to measure and record the movement characteristics under consideration [26–32, 35–54]. This approach represents the feedback portion in the context of the Internet of Things for establishing efficacy of Network Centric Therapy.

1.2.3 Perspective of Chap. 4: Deep Brain Stimulation for the Treatment of Movement Disorder Regarding Parkinson's Disease and Essential Tremor with Device Characterization

Deep brain stimulation offers a long-term treatment intervention for progressive movement disorders, such as Parkinson's disease and Essential tremor [11, 55–60]. Its origins are attributed to the research endeavors of Dr. Benabid on the order of a quarter of a century ago. Deep brain stimulation provides an alternative to conditions, for which medication becomes intractable without the permanency of lesioning neurosurgery [11, 55]. Risks also exist regarding the implantation and operation of a deep brain stimulation system, which are generally constrained to the surgical and stimulation-based adverse effects [59]. However, intrinsic advantages for deep brain stimulation are considerable, such as the reversible and even adjustable characteristics of the therapy [61].

The deep brain stimulation device is presented from a general systems perspective. Emphasis is placed on the implantable pulse generator, battery, connecting wire, and electrode lead [12]. Beyond the surgical environment, there exists considerable challenge for the expert utilization of the deep brain stimulation system, such as the objective of attaining an optimal parameter configuration [12–14]. On the order of thousands of permutations exist as multiple parameters can be varied, such as polarity, amplitude of stimulation, rate of stimulation, and pulse width [12]. Current strategies for converging upon an optimal parameter configuration are resource intensive and inherently tedious [62–64].

Future goals for the evolutionary development of deep brain stimulation involve the goal of achieving a closed-loop parameter tuning capability [65–68]. Preliminary breakthroughs for establishing this technology involve the application of wearable and wireless systems, such as a smartphone, for the objective and quantified feedback of movement disorder tremor status [35–37, 51]. Further advances involve the post-processing capabilities revealed through the application of machine learning to classify between disparate scenarios, such as for the deep brain stimulation system set to "On" and "Off" [35–37]. These developments amalgamate to establish the rampant presence of Network Centric Therapy, for which optimal intervention strategies are enabled. With Network Centric Therapy, the patient and clinical resources can reside in remote settings with continuous interaction provided by wireless connectivity to the Internet.

1.2.4 Perspective of Chap. 5: Surgical Procedure for Deep Brain Stimulation Implantation and Operative Phase with Postoperative Risks

The process for implantation of a deep brain stimulation system does comprise inherent risk. However, with respect to many scenarios, the benefit of ameliorating movement disorder symptoms, such as Parkinson's disease and Essential tremor, is relatively greater. A carefully planned surgical procedure is applied in a phased approach, for which some risks involve infection and cerebral hemorrhaging [69–71]. With regard to long-term application of deep brain stimulation, the presence of extrinsic electromagnetic fields can pose extreme danger to the patient [72, 73]. This observation underscores the imperative safety considerations for using magnetic resonance imaging to determine the quality and spatial positioning of the deep brain stimulation system electrodes [74, 75]. Adverse neurological and neuropsychological effects are also potential consequences of deep brain stimulation [69, 76]. An applied demonstration of expert surgical methodology for installing the deep brain stimulation system is presented, in conjunction with the expert perceptivity imperative for the optimization of parameter configuration tuning process [77]. This technique establishes the motivation for the prominence of wearable and wireless systems as a quantified feedback strategy for deep brain stimulation, which is foundational for the development of Network Centric Therapy for treating neurodegenerative movement disorders, such as Parkinson's disease and Essential tremor.

1.2.5   Perspective of Chap. 6: Preliminary Wearable and Locally Wireless Systems for Quantification of Parkinson's Disease and Essential Tremor Characteristics

The origins of Network Centric Therapy derive from the application of preliminary wearable and wireless systems with local range of connectivity. The application of inertial sensors for the quantification of human movement has been a concept under consideration for more than a half century [26, 32, 78]. With progressive technological evolution through the influence of extrinsic industries, these inertial sensors achieved the characteristics for integration into the biomedical community [26, 32, 79]. Eventually, inertial sensors attained wireless capabilities, which outmoded more cumbersome means of data transmission [80].

Preliminary research, development, testing, and evaluation of wearable and locally wireless systems demonstrated the opportunity to quantify Parkinson's disease hand tremor. Highly miniaturized wireless accelerometers were mounted about the dorsum of the hand and even secured by a functional glove to constitute a wearable system. From the wearable and wireless accelerometer system, the signal data would be conveyed by local wireless transmission to a proximally situated computer for post-processing [50, 53, 81, 82]. Logically this capability can be extrapolated to other types of neurodegenerative movement disorders, such as Essential tremor. Access to the Internet from the proximally situated computer that receives signal data from the local wireless accelerometer promotes a preliminary vantage regarding the future opportunities enabled by Network Centric Therapy, such as highly remote post-processing.

1.2.6   Perspective of Chap. 7: Wearable and Wireless Systems with Internet Connectivity for Quantification of Parkinson's Disease and Essential Tremor Characteristics

The previous Chap. 6 "Preliminary wearable and locally wireless systems for quantification of Parkinson's disease and Essential tremor characteristics" establishes the preliminary foundation of Network Centric Therapy. This chapter extrapolates the technology to the origins of Network Centric Therapy through the application of wearable and wireless systems for the quantification of Parkinson's disease tremor and Essential tremor. This type of wearable and wireless system bypasses the requirement for local wireless connectivity to a nearby situated computer for access to the Internet, since the wearable and wireless system, such as a smartphone, can access the Internet through wireless connectivity [27–31, 35–37, 51].

During 2010 LeMoyne et al. demonstrated the ability of the smartphone to quantify Parkinson's disease hand tremor. The smartphone was mounted to the dorsum of the hand through a glove. The hand tremor was quantified through the smartphone's accelerometer, and the signal data was conveyed by wireless transmission to the Internet as an email attachment. The experimental location and the post-processing resources were situated on effectively opposite sides of the continental United States [51]. The engineering proof of concept perspective provided by LeMoyne et al. during 2010 establishes the origins of Network Centric Therapy.

Further extension of this capability pertains to the machine learning classification of deep brain stimulation set to "On" and "Off" for both Parkinson's disease and Essential tremor. Software automation provides the opportunity to consolidate a feature set using the inertial sensor signal. Considerable classification accuracy for distinguishing between the deep brain stimulation set to "On" and "Off" for both Parkinson's disease and Essential tremor has been attained [36, 37].

Another extension of Network Centric Therapy for the evaluation of neurodegenerative movement disorders, such as Parkinson's disease and Essential tremor, is the development of locally wireless inertial sensor nodes that have connectivity to a smartphone or tablet. An advantage of this perspective is the capacity to further miniaturize the locally wireless inertial sensor nodes without the need for long-range connectivity. By transmitting the signal data to a smartphone or tablet, the data package can then be conveyed wirelessly to Cloud computing resources for post-processing [27, 83]. One of the most predominant forms of post-processing the signal data is through the application of machine learning classification [84].

1.2.7 Perspective of Chap. 8: Role of Machine Learning for Classification of Movement Disorder and Deep Brain Stimulation Status

Integrating the concepts of machine learning in conjunction with wearable and wireless systems for the domain of neurodegenerative movement disorders, such as Parkinson's disease and Essential tremor, may offer a quantum leap relative to conventional diagnostic methodologies. Machine learning algorithms are envisioned to enable the determination of efficacy for medical therapy strategies and the derivation of optimal parameter configurations for deep brain stimulation. This opportunity is provided by the inertial signal data acquired by the wearable and wireless system [26–31]. The concept of applying machine learning for neurodegenerative movement disorders, such as Parkinson's disease and Essential tremor, has been proposed and successfully implemented with respect to attaining considerable classification accuracy to differentiate between deep brain stimulation set to "On" and "Off" status [35–37, 85].

The software platform that is central to these machine learning endeavors is the Waikato Environment for Knowledge Analysis (WEKA). WEKA consists of a considerable assortment of machine learning algorithms. Six machine learning algorithms are emphasized:

- J48 decision tree
- K-nearest neighbors
- Logistic regression
- Support vector machine
- Multilayer perceptron neural network
- Random forest [86–88]

The foundation to successful application of machine learning is the acquisition of an appropriate feature set, which is generally acquired through the use a software automation process. With regard to the integration of wearable and wireless systems for neurodegenerative movement disorders, such as Parkinson's disease and Essential tremor, the inertial sensor signal is consolidated into an Attribute-Relation File Format (ARFF) for ascertaining machine learning classification with WEKA [35–37, 85–88].

Machine learning is envisioned to enable considerable diagnostic and prognostic capabilities for Network Centric Therapy with respect to post-processing from the Cloud computing resource. In light of the considerable evolution of computational power for wearable and wireless systems, machine learning classification techniques may also be applied prior to wireless data transmission to the Cloud computing resource from the wearable and wireless system itself. This opportunity could facilitate the allocation of resources and hasten the interactive process for optimal treatment of neurodegenerative movement disorders, such as Parkinson's disease and Essential tremor.

1.2.8   Perspective of Chap. 9: Assessment of Machine Learning Classification Strategies for the Differentiation of Deep Brain Stimulation "On" and "Off" Status for Parkinson's Disease Using a Smartphone as a Wearable and Wireless Inertial Sensor for Quantified Feedback

The synergistic utility of amalgamating deep brain stimulation using a smartphone as a wearable and wireless inertial sensor system with machine learning for establishing classification accuracy is presented. The appropriateness of a machine learning strategy is highly associated with the intrinsic characteristics of the feature set selected for classification [89]. Therefore, the consideration of multiple machine learning classification strategies would provide long-term benefit to the progressive development of Network Centric Therapy for treating progressive neurodegenerative movement disorders, such as Parkinson's disease and Essential tremor.

Using Waikato Environment for Knowledge Analysis (WEKA), six machine learning algorithms have been selected for the distinguishing between deep brain stimulation set to "On" and "Off" status:

- J48 decision tree
- K-nearest neighbors
- Logistic regression
- Support vector machine
- Multilayer perceptron neural network
- Random forest [86–88]

The machine learning classification accuracy differentiating between deep brain stimulation set to "On" and "Off" status using a smartphone is based on successfully presented research endeavors [37, 90]. Within the scope of the selected machine learning strategies, considerable classification accuracy is attained. Regarding the machine learning algorithms achieving the highest classification accuracy, their processing time is also discussed for greater consideration of computational efficiency.

The dual benefits of classification accuracy and computational efficiency provide insight regarding the architectural perspective for Network Centric Therapy. The machine learning endeavor could be conducted locally to the wearable and wireless inertial sensor or using the Cloud-situated post-processing resources. The selection of the proper architectural perspective could be based upon the appropriate response time for the acquisition of an optimal deep brain stimulation system parameter configuration.

1.2.9 Perspective of Chap. 10: New Perspectives for Network Centric Therapy for the Treatment of Parkinson's Disease and Essential Tremor

The historic and current status of wearable and wireless systems for the treatment of progressive neurodegenerative movement disorders, such as Parkinson's disease and Essential tremor, using intervention techniques, such as deep brain stimulation, has been thoroughly established. As an extension of this capability, perspective future extrapolations to the domain of Network Centric Therapy are addressed. With wearable and wireless inertial sensors establishing Cloud computing historic databases, a new era of data science for the robust treatment of progressive neurodegenerative movement disorders, such as Parkinson's disease and Essential tremor, is established. Optimal deep brain stimulation configuration parameters can be acquired effectively in real-time for providing efficacious treatment. In addition, the global nature of the Internet facilitates therapy intervention with the best clinical resources anywhere in the world.

1.3 Conclusion

The benefit of wearable and wireless inertial sensor systems, such as the smartphone, for the objective quantification of movement disorder tremor, such as Parkinson's disease and Essential tremor, has been thoroughly advocated. Wearable and wireless inertial sensor systems have progressively evolved for enhanced suitability regarding the evaluation of human movement characteristics for clinical application. In particular wearable and wireless inertial sensor systems constitute a transcendent technology when contrasted to conventional ordinal scale techniques for diagnosing the severity of progressive neurodegenerative movement disorders, such as Parkinson's disease and Essential tremor.

Synergy with the evolution of wearable and wireless inertial sensor systems, such as the smartphone, is the advent of the deep brain stimulation system for ameliorating tremor for Parkinson's disease and Essential tremor. However, even for an expertly skilled clinician, the task of converging upon an efficacious parameter configuration for a deep brain stimulation system can present a daunting endeavor. The quantified feedback enabled by a wearable and wireless inertial sensor system can provide objective insight as to the efficacy of a deep brain stimulation system parameter configuration. For example, with the further amalgamation of machine learning, deep brain stimulation "On" and "Off" status has been distinguished with considerable classification accuracy using inertial sensor signal data from a wearable and wireless system to compose the respective feature set. Since the wearable and wireless system, such as a smartphone, can convey the experimental trial data by wireless connectivity to the Internet as an email attachment, the experimental and post-processing resources can be remotely situated anywhere in the world.

The capacity to wirelessly access Internet resources with wearable and wireless systems, such as a smartphone, establishes the foundation for Network Centric Therapy for the treatment of progressive neurodegenerative movement disorders, such as Parkinson's disease and Essential tremor using the intervention of deep brain stimulation systems. In the near future, Network Centric Therapy should be capable of facilitating real-time optimization of deep brain stimulation configuration parameters through the objectively quantified status of progressive neurodegenerative movement disorders, such as Parkinson's disease and Essential tremor, using inertial sensor signal data acquired by a wearable and wireless system, such as a smartphone. These implications are envisioned to provide people with progressive neurodegenerative movement disorders, such as Parkinson's disease and Essential tremor, global access to expert clinical resources with optimal therapy intervention strategies.

References

1. Parkinson J (1817) An essay on the shaking palsy. Whittingham and Rowland, London
2. Louis ED (2005) Essential tremor. Lancet Neurol 4(2):100–110
3. Louis ED (2000) Essential tremor. Arch Neurol (JAMA Neurology) 57(10):1522–1524
4. Kandel ER, Schwartz JH, Jessell TM (2000) Principles of neural science. McGraw-Hill, New York, Ch 43
5. Essential tremor: [http://www.essentialtremor.org/about-et/]
6. Seeley RR, Stephens TD, Tate P (2003) Anatomy and physiology. McGraw-Hill, Boston, Ch 14
7. Deuschl G, Raethjen J, Hellriegel H, Elble R (2011) Treatment of patients with essential tremor. Lancet Neurol 10(2):148–161
8. Habib-ur-Rehman (2000) Diagnosis and management of tremor. Arch Intern Med 160(16):2438–2444
9. LeMoyne R (2013) Wearable and wireless accelerometer systems for monitoring Parkinson's disease patients—a perspective review. Adv Park Dis 2(4):113–115
10. Nolte J, Sundsten JW (2002) The human brain: an introduction to its functional anatomy. Mosby, St. Louis, Ch 19
11. Williams R (2010) Alim-Louis Benabid: stimulation and serendipity. Lancet Neurol 9(12):1152
12. Amon A, Alesch F (2017) Systems for deep brain stimulation: review of technical features. J Neural Transm 124(9):1083–1091
13. Isaias IU, Tagliati M (2008) Deep brain stimulation programming for movement disorders. In: Deep brain stimulation in neurological and psychiatric disorders. Springer, New York, pp 361–397
14. Volkmann J, Moro E, Pahwa R (2006) Basic algorithms for the programming of deep brain stimulation in Parkinson's disease. Mov Disord 21(S14):S284–S289
15. Ramaker C, Marinus J, Stiggelbout AM, Van Hilten BJ (2002) Systematic evaluation of rating scales for impairment and disability in Parkinson's disease. Mov Disord 17(5):867–876
16. Fahn S, Elton RL, UPDRS Program Members (1987) Unified Parkinson's Disease Rating Scale. In: Recent developments in Parkinson's disease, Vol. 2. Macmillan Healthcare Information, Florham Park, pp 153–163, 293–304.
17. Goetz CG, Stebbins GT, Chmura TA, Fahn S, Poewe W, Tanner CM (2010) Teaching program for the Movement Disorder Society-sponsored revision of the Unified Parkinson's Disease Rating Scale: (MDS-UPDRS). Mov Disord 25(9):1190–1194
18. Movement Disorder Society Task Force on Rating Scales for Parkinson's Disease (2003) The Unified Parkinson's Disease Rating Scale (UPDRS): status and recommendations. Mov Disord 18(7):738–750
19. Goetz CG, Tilley BC, Shaftman SR, Stebbins GT, Fahn S, Martinez-Martin P, Poewe W, Sampaio C, Stern MB, Dodel R, Dubois B, Holloway R, Jankovic J, Kulisevsky J, Lang AE, Lees A, Leurgans S, LeWitt PA, Nyenhuis D, Olanow CW, Rascol O, Schrag A, Teresi JA, van Hilten JJ, LaPelle N (2008) Movement Disorder Society-sponsored revision of the Unified Parkinson's Disease Rating Scale (MDS-UPDRS): scale presentation and clinimetric testing results. Mov Disord 23(15):2129–2170
20. Fahn S, Tolosa E, Marin C (1988) Clinical rating scale for tremor. In: Parkinson's disease and movement disorders. Urban & Schwarzenberg, Baltimore, pp 225–234
21. Elble RJ (2016) The essential tremor rating assessment scale. J Neurol Neuromed 1(4):34–38
22. Siderowf A, McDermott M, Kieburtz K, Blindauer K, Plumb S, Shoulson I (2002) Test–retest reliability of the unified Parkinson's disease rating scale in patients with early Parkinson's disease: results from a multicenter clinical trial. Mov Disord 17(4):758–763
23. Metman LV, Myre B, Verwey N, Hassin-Baer S, Arzbaecher J, Sierens D, Bakay R (2004) Test–retest reliability of UPDRS-III, dyskinesia scales, and timed motor tests in patients with advanced Parkinson's disease: an argument against multiple baseline assessments. Mov Disord 19(9):1079–1084

24. Richards M, Marder K, Cote L, Mayeux R (1994) Interrater reliability of the Unified Parkinson's Disease Rating Scale motor examination. Mov Disord 9(1):89–91
25. Post B, Merkus MP, de Bie RM, de Haan RJ, Speelman JD (2005) Unified Parkinson's Disease Rating Scale motor examination: are ratings of nurses, residents in neurology, and movement disorders specialists interchangeable? Mov Disord 20(12):1577–1584
26. LeMoyne R, Coroian C, Cozza M, Opalinski P, Mastroianni T, Grundfest W (2009) The merits of artificial proprioception, with applications in biofeedback gait rehabilitation concepts and movement disorder characterization. In: Biomedical engineering. InTech, Vienna, pp 165–198
27. LeMoyne R, Mastroianni T (2018) Wearable and wireless systems for healthcare I: gait and reflex response quantification. Springer, Singapore
28. LeMoyne R, Mastroianni T (2017) Smartphone and portable media device: a novel pathway toward the diagnostic characterization of human movement. In: Smartphones from an applied research perspective. InTech, Rijeka, Croatia, pp 1–24
29. LeMoyne R, Mastroianni T (2017) Wearable and wireless gait analysis platforms: smartphones and portable media devices. In: Wireless MEMS networks and applications. Elsevier, New York, pp 129–152
30. LeMoyne R, Mastroianni T (2016) Telemedicine perspectives for wearable and wireless applications serving the domain of neurorehabilitation and movement disorder treatment. In: Telemedicine, SMGroup, Dover, Delaware, pp 1–10
31. LeMoyne R, Mastroianni T (2015) Use of smartphones and portable media devices for quantifying human movement characteristics of gait, tendon reflex response, and Parkinson's disease hand tremor. In: Mobile health technologies, methods and protocols. Springer, New York, pp 335–358
32. LeMoyne R, Coroian C, Mastroianni T, Grundfest W (2008) Accelerometers for quantification of gait and movement disorders: a perspective review. J Mech Med Biol 8(2):137–152
33. Diamond MC, Scheibel AB, Elson LM (1985) The human brain coloring book. Harper Perennial, New York, Ch 5
34. Hariz GM, Lindberg M, Bergenheim AT (2002) Impact of thalamic deep brain stimulation on disability and health-related quality of life in patients with essential tremor. J Neurol Neurosurg Psychiatry 72(1):47–52
35. LeMoyne R, Tomycz N, Mastroianni T, McCandless C, Cozza M, Peduto D (2015) Implementation of a smartphone wireless accelerometer platform for establishing deep brain stimulation treatment efficacy of essential tremor with machine learning. In: 37th Annual international conference of the IEEE, Engineering in Medicine and Biology Society (EMBS), pp 6772–6775
36. LeMoyne R, Mastroianni T, Tomycz N, Whiting D, Oh M, McCandless C, Currivan C, Peduto D (2017) Implementation of a multilayer perceptron neural network for classifying deep brain stimulation in 'On' and 'Off' modes through a smartphone representing a wearable and wireless sensor application. In: 47th Society for Neuroscience annual meeting (featured in Hot Topics; top 1% of abstracts)
37. LeMoyne R, Mastroianni T, McCandless C, Currivan C, Whiting D, Tomycz N (2018) Implementation of a smartphone as a wearable and wireless accelerometer and gyroscope platform for ascertaining deep brain stimulation treatment efficacy of Parkinson's disease through machine learning classification. Adv Park Dis 7(2):19–30
38. LeMoyne RC (2010) Wireless quantified reflex device. Ph.D. Dissertation UCLA
39. LeMoyne R, Mastroianni T, Coroian C, Grundfest W (2011) Tendon reflex and strategies for quantification, with novel methods incorporating wireless accelerometer reflex quantification devices, a perspective review. J Mech Med Biol 11(3):471–513
40. LeMoyne R, Mastroianni T, Kale H, Luna J, Stewart J, Elliot S, Bryan F, Coroian C, Grundfest W (2011) Fourth generation wireless reflex quantification system for acquiring tendon reflex response and latency. J Mech Med Biol 11(1):31–54
41. LeMoyne R, Coroian C, Mastroianni T, Grundfest W (2008) Quantified deep tendon reflex device for response and latency, third generation. J Mech Med Biol 8(4):491–506

42. LeMoyne R, Dabiri F, Jafari R (2008) Quantified deep tendon reflex device, second generation. J Mech Med Biol 8(1):75–85
43. LeMoyne R, Dabiri F, Coroian C, Mastroianni T, Grundfest W (2007) Quantified deep tendon reflex device for assessing response and latency. In: 37th Society for Neuroscience annual meeting
44. LeMoyne R, Jafari R, Jea D (2005) Fully quantified evaluation of myotatic stretch reflex. In: 35th Society for Neuroscience annual meeting
45. LeMoyne R, Mastroianni T (2017) Implementation of a smartphone wireless gyroscope platform with machine learning for classifying disparity of a hemiplegic patellar tendon reflex pair. J Mech Med Biol 17(6):1750083
46. LeMoyne R, Kerr W, Zanjani K, Mastroianni T (2014) Implementation of an iPod wireless accelerometer application using machine learning to classify disparity of hemiplegic and healthy patellar tendon reflex pair. J Med Imaging Health Inform 4(1):21–28
47. LeMoyne R, Mastroianni T (2016) Smartphone wireless gyroscope platform for machine learning classification of hemiplegic patellar tendon reflex pair disparity through a multilayer perceptron neural network. In: Wireless Health (WH) of IEEE, pp 103–108
48. LeMoyne R, Mastroianni T (2015) Machine learning classification of a hemiplegic and healthy patellar tendon reflex pair through an iPod wireless gyroscope platform. In: 45th Society for Neuroscience annual meeting
49. LeMoyne R, Mastroianni T, Grundfest W, Nishikawa K (2013) Implementation of an iPhone wireless accelerometer application for the quantification of reflex response. In: 35th Annual international conference of the IEEE, Engineering in Medicine and Biology Society (EMBS), pp 4658–4661
50. LeMoyne R, Coroian C, Mastroianni T (2009) Quantification of Parkinson's disease characteristics using wireless accelerometers. In: ICME International conference on IEEE Complex Medical Engineering (CME), pp 1–5
51. LeMoyne R, Mastroianni T, Cozza M, Coroian C, Grundfest W (2010) Implementation of an iPhone for characterizing Parkinson's disease tremor through a wireless accelerometer application. In: 32nd Annual international conference of the IEEE, Engineering in Medicine and Biology Society (EMBS), pp 4954–4958
52. LeMoyne R, Mastroianni T (2016) Implementation of a multilayer perceptron neural network for classifying a hemiplegic and healthy reflex pair using an iPod wireless gyroscope platform. In: 46th Society for Neuroscience annual meeting
53. LeMoyne R, Mastroianni T, Grundfest W (2013) Wireless accelerometer configuration for monitoring Parkinson's disease hand tremor. Adv Park Dis 2(2):62–67
54. LeMoyne R, Coroian C, Mastroianni T, Grundfest W (2008) Virtual proprioception. J Mech Med Biol 8(3):317–338
55. Benabid AL, Pollak P, Louveau A, Henry S, de Rougemont J (1987) Combined (thalamotomy and stimulation) stereotactic surgery of the VIM thalamic nucleus for bilateral Parkinson's disease. Appl Neurophysiol 50(1–6):344–346
56. Rehncrona S, Johnels B, Widner H, Törnqvist AL, Hariz M, Sydow O (2003) Long-term efficacy of thalamic deep brain stimulation for tremor: double-blind assessments. Mov Disord 18(2):163–170
57. Sydow O, Thobois S, Alesch F, Speelman JD (2003) Multicentre European study of thalamic stimulation in essential tremor: a six year follow up. J Neurol Neurosurg Psychiatry 74(10):1387–1391
58. Krack P, Batir A, Van Blercom N, Chabardes S, Fraix V, Ardouin C, Koudsie A, Limousin PD, Benazzouz A, LeBas JF, Benabid AL, Pollak P (2003) Five-year follow-up of bilateral stimulation of the subthalamic nucleus in advanced Parkinson's disease. N Engl J Med 349(20):1925–1934
59. Lyons KE, Koller WC, Wilkinson SB, Pahwa R (2001) Long term safety and efficacy of unilateral deep brain stimulation of the thalamus for parkinsonian tremor. J Neurol Neurosurg Psychiatry 71(5):682–684

60. Benabid AL, Benazzous A, Pollak P (2002) Mechanisms of deep brain stimulation. Mov Disord 17(S3):S73–S74
61. Yu H, Neimat JS (2008) The treatment of movement disorders by deep brain stimulation. Neurotherapeutics 5(1):26–36
62. Pretto T (2007) Deep brain stimulation. Neurologist 13(2):103–104
63. Panisset M, Picillo M, Jodoin N, Poon YY, Valencia-Mizrachi A, Fasano A, Munhoz R, Honey CR (2017) Establishing a standard of care for deep brain stimulation centers in Canada. Can J Neurol Sci 44(2):132–138
64. Schwalb JM, Hamani C (2008) The history and future of deep brain stimulation. Neurotherapeutics 5(1):3–13
65. Hariz M (2017) My 25 stimulating years with DBS in Parkinson's disease. J Park Dis 7(s1):S33–S41
66. Fang JY, Tolleson C (2017) The role of deep brain stimulation in Parkinson's disease: an overview and update on new developments. Neuropsychiatr Dis Treat 13:723–732
67. Sun FT, Morrell MJ (2014) Closed-loop neurostimulation: the clinical experience. Neurotherapeutics 11(3):553–563
68. Priori A, Foffani G, Rossi L, Marceglia S (2013) Adaptive deep brain stimulation (aDBS) controlled by local field potential oscillations. Exp Neurol 245:77–86
69. Okun MS (2012) Deep-brain stimulation for Parkinson's disease. N Engl J Med 367(16):1529–1538
70. Hariz MI (2002) Complications of deep brain stimulation surgery. Mov Disord 17(S3):S162–S166
71. Constantoyannis C, Berk C, Honey CR, Mendez I, Brownstone RM (2005) Reducing hardware-related complications of deep brain stimulation. Can J Neurol Sci 32(2):194–200
72. Patterson T, Stecker MM, Netherton BL (2007) Mechanisms of electrode induced injury. Part 2: clinical experience. Am J Electroneurodiagnostic Technol 47(2):93–113
73. Nutt JG, Anderson VC, Peacock JH, Hammerstad JP, Burchiel KJ (2001) DBS and diathermy interaction induces severe CNS damage. Neurology 56(10):1384–1386
74. Rezai AR, Phillips M, Baker KB, Sharan AD, Nyenhuis J, Tkach J, Henderson J, Shellock FG (2004) Neurostimulation system used for deep brain stimulation (DBS): MR safety issues and implications of failing to follow safety recommendations. Investig Radiol 39(5):300–303
75. Tagliati M, Jankovic J, Pagan F, Susatia F, Isaias IU, Okun MS (2009) Safety of MRI in patients with implanted deep brain stimulation devices. NeuroImage 47(S2):T53–T57
76. Temel Y (2010) Limbic effects of high-frequency stimulation of the subthalamic nucleus. Vitam Horm 82:47–63
77. Tomycz ND, Whiting DM (2018) Deep brain stimulation: indications, operative technique, and programming. Internal Publication Allegheny General Hospital
78. Saunders JB, Inman VT, Eberhart HD (1953) The major determinants in normal and pathological gait. J Bone Joint Surg 35A(3):543–558
79. Culhane KM, O'Connor M, Lyons D, Lyons GM (2005) Accelerometers in rehabilitation medicine for older adults. Age Ageing 34(6):556–560
80. Patel S, Park H, Bonato P, Chan L, Rodgers M (2012) A review of wearable sensors and systems with application in rehabilitation. J Neuroeng Rehabil 9(1):21
81. LeMoyne R (2007) Gradient optimized neuromodulation for Parkinson's disease. In: 12th Annual UCLA research conference on aging
82. LeMoyne R, Coroian C, Mastroianni T (2008) 3D wireless accelerometer characterization of Parkinson's disease status. In: Plasticity and repair in neurodegenerative disorders (Conference)
83. LeMoyne R, Mastroianni T (2018) Bluetooth inertial sensors for gait and reflex response quantification with perspectives regarding cloud computing and the Internet of Things. In: Wearable and wireless systems for healthcare I: gait and reflex response quantification. Springer, Singapore, pp 95–103

84. LeMoyne R, Mastroianni T (2018) Role of machine learning for gait and reflex response classification. In: Wearable and wireless systems for healthcare I: gait and reflex response quantification. Springer, Singapore, pp 111–120
85. LeMoyne R, Mastroianni T, Tomycz N, Whiting D, McCandless C, Peduto D, Cozza M (2015) I-Phone wireless accelerometer quantification of extremity tremor in essential tremor patient undergoing activated and inactivated deep brain stimulation. In: International Neuromodulation Society's 12th World Congress
86. Hall M, Frank E, Holmes G, Pfahringer B, Reutemann P, Witten IH (2009) The WEKA data mining software: an update. ACM SIGKDD Explor Newsl 11(1):10–18
87. Witten IH, Frank E, Hall MA (2011) Data mining: practical machine learning tools and techniques. Morgan Kaufmann, Burlington, MA
88. WEKA [http://www.cs.waikato.ac.nz/~ml/weka/]
89. LeMoyne R, Kerr W, Mastroianni T, Hessel A (2014) Implementation of machine learning for classifying hemiplegic gait disparity through use of a force plate. In: 13th International Conference on Machine Learning and Applications (ICMLA), IEEE, pp 379–382
90. LeMoyne R, Mastroianni T, McCandless C, Currivan C, Whiting D, Tomycz N (2018) Implementation of a smartphone as a wearable and wireless inertial sensor platform for determining efficacy of deep brain stimulation for Parkinson's disease tremor through machine learning. In: 48th Society for Neuroscience annual meeting (Nanosymposium)

Chapter 2
Movement Disorders: Parkinson's Disease and Essential Tremor—A General Perspective

Abstract Movement disorders manifesting in tremor influence the quality of life for millions of people. In particular, two prevalent types of movement disorder are Parkinson's disease and Essential tremor. The neurological foundation for Parkinson's disease is attributed to dysfunction of the substantia nigra and associated aspects of the basal ganglia. By contrast, Essential tremor is not conclusively defined. However, notable amplified cerebellar activity is a characteristic for Essential tremor. Traditional strategies for diagnosing the severity of Parkinson's disease and Essential tremor apply expert clinical although subjective interpretation of ordinal scales. This ordinal scale approach is the subject of contention regarding reliability. Traditional therapy involves the prescription of medication. As a last resort, permanent disruption of the deep brain neural pathways is an alternative. Recent developments have demonstrated the utility of wearable and wireless systems for the objective and quantified measurement of tremor symptoms. Furthermore, wearable and wireless systems have been amalgamated with deep brain stimulation for the determination of therapy efficacy. Near-term future objectives implicate the opportunity for real-time patient-specific optimization of deep brain stimulation tuning parameters. These developments lead to the presence of Network Centric Therapy for the treatment of movement disorders, such as Parkinson's disease and Essential tremor.

Keywords Movement disorder · Parkinson's disease · Essential tremor · Unified Parkinson's Disease Rating Scale (UPDRS) · Movement Disorder Society-Unified Parkinson's Disease Rating Scale (MDS-UPDRS) · Conventional medical intervention · Levodopa · Propranolol · Pallidotomy · Thalamotomy · Wearable and wireless systems · Deep brain stimulation · Network Centric Therapy

2.1 Introduction

Two prevalent movement disorders are addressed throughout the subject matter of the book: Parkinson's disease and Essential tremor. Both of these neurodegenerative diseases can be treated by medication and surgical intervention [1, 2]. However, these two types of movement disorders are attributed to different pathophysiological mechanisms [3]. Although Parkinson's disease and Essential tremor may appear highly similar to the common observer, their disparities are significant enough to warrant consideration, especially in the context of traditional clinical evaluation and intervention.

2.2 Parkinson's Disease

Dr. James Parkinson authored the seminal publication "An essay on the shaking palsy" during 1817. The publication thoroughly characterizes shaking palsy; a neurodegenerative disorder that is now attributed to his name [4]. The incidence of Parkinson's disease is generally prevalent for people exceeding 55 years of age [5]. Parkinson's disease is a medical issue of concern as on the order of one million people in the United States are diagnosed with the neurodegenerative disorder [1].

There are four standard neuromotor characteristics that facilitate the diagnosis of Parkinson's disease:

- Balance impairment
- Shuffling of gait
- Rigidity which is attributed to increase in muscle tone
- Tremor that is observable during resting status [1]

A notable observation is that resting tremor may decrease in severity or even disappear when the subject with Parkinson's disease conducts voluntary movement [6]. The frequency of Parkinson's disease resting tremor is bound between approximately four and five per second [1, 6].

The neurological origins of Parkinson's disease have been established. The degeneration of dopaminergic neurons in the substantia nigra is associated with the progressive onset of Parkinson's disease [1]. The symptoms of Parkinson's disease develop with the continual depletion of dopamine production for the putamen and caudate. These neurological structures comprise the basal ganglia [7].

2.3 Essential Tremor

Over the duration of approximately two millennia, medical intervention strategies have been developed for the objective of ameliorating symptoms, such as kinetic tremor [8, 9]. By relative comparison the incidence of Essential tremor is greater than for Parkinson's disease. In the United States, an estimated 10 million people have Essential tremor [10]. The occurrence of Essential tremor is proportional to age. The typical symptoms of Essential tremor most commonly manifest in the later 40s. However, children have been known to exhibit the symptoms of Essential tremor [8, 10]. The incidence of this neurodegenerative movement disorder exceeds the incidence of Parkinson's disease by an order of magnitude.

Although Essential tremor and Parkinson's disease both manifest movement disorder tremor symptoms, their tremor characteristics are notably different [1, 6, 8]. Essential tremor commonly displays kinetic tremor instead of resting tremor. Intuitively such tremor resulting from intentional movement may significantly influence daily life activities. These kinetic tremors are bound between a frequency range of 4–12 Hz [8]. Note that Essential tremor involves a higher frequency that transcends the frequency bounds of Parkinson's disease.

The fundamental neurological mechanisms that lead to the symptoms of Essential tremor have not been conclusively determined, which challenges the efficacy of an intervention strategy [8, 11]. Patients diagnosed with Essential tremor have demonstrated amplified cerebellar activity, as shown by positron emission tomography [12]. A research study by Paris-Robidas et al. advocate that the dentate nucleus experiences a decrement of GABA receptors. This decrement leads to the output activity for cerebellum pacemakers to become decremented respective of their inhibitive properties, which emanate through the cerebello-thalamo-cortical tract causing the occurrence of tremors [13].

2.4   Traditional Strategies for Assessing Progression and Treatment of Parkinson's Disease and Essential Tremor

Parkinson's disease and Essential tremor are both currently evaluated for status of progression through the application of ordinal status methodologies [14–16]. The Unified Parkinson's Disease Rating Scale (UPDRS) and evolved Movement Disorder Society-Unified Parkinson's Disease Rating Scale (MDS-UPDRS) emphasize both cognitive- and motor-related aspects of the impact of Parkinson's disease respective of the patient [14, 17, 18]. By contrast the Fahn-Tolosa-Marin Tremor Rating Scale for evaluating Essential tremor primarily emphasizes the nature of tremor and influence on motor tasks [15, 16]. Both of these evaluation techniques for Parkinson's disease and Essential tremor consist of a five-point ordinal scale from 0 to 4 with the increase in ordinal magnitude in direct proportion to symptom severity [14–16].

Intuitively, since these ordinal scale techniques involve the evaluation of disparate symptoms, medication used for treatment intervention is likewise generally disparate [12]. The administration of Levodopa is a prevalent medical intervention for the regulation of Parkinson's disease tremor symptoms [1, 12, 19]. Propranolol is a standard drug for the treatment of Essential tremor [8, 11]. Propranolol is also relevant as a drug therapy for Parkinson's disease. There are multiple other medication strategies for the treatment of Parkinson's disease and Essential tremor [12].

From the traditional therapy strategy perspective in the event that medication becomes intractable for resolving movement disorder tremor, permanent disruption of the neural pathways of the deep brain is an option of last resort [1, 2, 8]. Two standard neurological procedures requiring high-precision stereotactic techniques for the treatment of the Parkinson's disease are the pallidotomy and thalamotomy [1, 2, 20–22]. The pallidotomy involves disrupting the internal segment of the globus pallidus [1, 2]. The thalamotomy disrupts the thalamus, for which such thermal coagulation techniques can be implemented through devices, such as the gamma knife [1, 2, 21, 22].

2.5 Advanced Strategies for Assessing Progression and Treatment of Parkinson's Disease and Essential Tremor

Parkinson's disease and Essential tremor have been noted for their distinctive tremor frequencies [1, 6, 8]. Intuitively, a measurement device, such as an inertial sensor system, may classify between the two movement disorders. The application of an accelerometer to distinguish between Parkinson's disease tremor and Essential tremor has been demonstrated by Hossen et al. [23]. LeMoyne et al. have demonstrated the quantification of Parkinson's disease tremor and Essential tremor though wearable and wireless systems comprised of inertial sensors, such as an accelerometer and gyroscope, for the objective assessment of movement disorder status and even for feedback of deep brain stimulation efficacy [19, 24–33].

Deep brain stimulation constitutes a highly novel therapy intervention for the treatment of Parkinson's disease and Essential tremor [34–37]. The deep brain stimulation system can be tuned with parameters, such as amplitude, frequency, pulse width, and electrode polarity [36]. Ascertaining an optimal parameter configuration for even an expert clinician can prove to be a daunting endeavor that is inherently time consuming [38]. A near-term future research objective is to combine the opportunities of wearable and wireless systems with deep brain stimulation systems for the development of effectively real-time determination of patient optimized parameter configurations. This near-term future research objective enables the development of Network Centric Therapy for movement disorders, such as Parkinson's disease and Essential tremor.

2.6 Extrapolation to Network Centric Therapy

Network Centric Therapy is effectively a representation of the Internet of Things for the domain of healthcare. In particular, the scope of this book is pertinent to the presence of Network Centric Therapy for the domains of Parkinson's disease and Essential tremor. Network Centric Therapy is a sharp contrast to the traditional approach of diagnosing and treating movement disorders. Rather than retaining paper documentation of subjective ordinal scale strategy evaluations, Network Centric Therapy offers the opportunity to store inertial signal data of tremor status in Cloud computing databases through objective and quantified wearable and wireless systems. Historical patient-specific Cloud computing databases can be applied to transformative data science techniques. Wearable and wireless systems for objectively quantifying tremor symptoms regarding Parkinson's disease and Essential tremor would remotely connect patient and the clinical therapy team from any location in the world by means of the Internet.

Network Centric Therapy offers the opportunity to determine the efficacy of deep brain stimulation through the feedback of wearable and wireless systems. Parameter configurations for deep brain stimulation can be upgraded by the clinician via Internet connectivity upon considering the inertial signal data of the wearable and wireless systems. Not only is historical status tracking of the neurodegenerative progression of Parkinson's disease and Essential tremor feasible, but closed-loop optimization of the deep brain stimulation system through real-time feedback derived from wearable and wireless inertial sensor systems is a highly realistic opportunity.

2.7 Conclusion

Parkinson's disease and Essential tremor are two highly prevalent types of neurodegenerative movement disorder that represent the scope of this book. On the scale of centuries, both Parkinson's disease and Essential tremor have been successfully diagnosed. Although to a layman the symptoms of these two kinds of movement disorder may appear to be quite similar, the expert clinician can readily distinguish between Parkinson's disease and Essential tremor. For example, the tremor frequency range of Parkinson's disease and Essential tremor is perceptibly disparate.

Although Parkinson's disease and Essential tremor are both currently evaluated by traditional ordinal scale methodologies, different ordinal scales are applied. The Unified Parkinson's Disease Rating Scale (UPDRS) and evolved Movement Disorder Society-Unified Parkinson's Disease Rating Scale (MDS-UPDRS) are utilized for Parkinson's disease and Fahn-Tolosa-Marin Tremor Rating Scale pertains to Essential tremor. The neurological foundation of Parkinson's disease is attributed to dysfunction of the substantia nigra. However, the neurological origins for developing Essential tremor are not currently established.

Treatment of Parkinson's disease and Essential tremor is traditionally accomplished through intervention by medicine, such as Levodopa for Parkinson's disease and Propranolol for Essential tremor. In case that a medication strategy becomes inefficacious, disruption of deep brain pathways is an ultimate contingency, such as through a pallidotomy or thalamotomy. In recent decades deep brain stimulation offers a breakthrough therapy alternative. Ascertaining the optimal parameter configuration is a challenging endeavor even for a skilled clinician.

Wearable and wireless inertial sensor systems offer another opportunity for the optimal treatment of Parkinson's disease and Essential tremor. Recently, wearable and wireless inertial sensor systems have been demonstrated for the determination of deep brain stimulation system efficacy. These developments advocate the development of Network Centric Therapy for the treatment of Parkinson's disease and Essential tremor. Network Centric Therapy represents the Internet of Things for the objective quantification of neurodegenerative movement disorder status with historical database tracking, which provides a basis for data science to resolve Parkinson's disease and Essential tremor. Furthermore, the integration of wearable and wireless inertial sensor systems with deep brain stimulation systems facilitates the ability to optimize unique parameter configurations according to the needs of the patient eventually in a real-time context.

References

1. Kandel ER, Schwartz JH, Jessell TM (2000) Principles of neural science. McGraw-Hill, New York, Ch 43
2. Nolte J, Sundsten JW (2002) The human brain: an introduction to its functional anatomy, St. Louis, Mosby, Ch 19
3. Helmich RC, Toni I, Deuschl G, Bloem BR (2013) The pathophysiology of essential tremor and Parkinson's tremor. Curr Neurol Neurosci Rep 13(9):378
4. Parkinson J (1817) An essay on the shaking palsy. Whittingham and Rowland, London
5. Seeley RR, Stephens TD, Tate P (2003) Anatomy and physiology. McGraw-Hill, Boston, Ch 14
6. Bickley LS, Szilagyi PG (2003) Bates' guide to physical examination and history taking. Lippincott Williams and Wilkins, Philadelphia, Ch 16
7. Diamond MC, Scheibel AB, Elson LM (1985) The human brain coloring book. Harper Perennial, New York, Ch 5
8. Louis ED (2005) Essential tremor. Lancet Neurol 4(2):100–110
9. Louis ED (2000) Essential tremor. Arch Neurol (JAMA Neurology) 57(10):1522–1524
10. Essential tremor: [http://www.essentialtremor.org/about-et/]
11. Deuschl G, Raethjen J, Hellriegel H, Elble R (2011) Treatment of patients with essential tremor. Lancet Neurol 10(2):148–161
12. Habib-ur-Rehman (2000) Diagnosis and management of tremor. Arch Intern Med 160(16):2438–2444
13. Paris-Robidas S, Brochu E, Sintes M, Emond V, Bousquet M, Vandal M, Pilote M, Tremblay C, Di Paolo T, Rajput AH, Rajput A, Calon F (2012) Defective dentate nucleus GABA receptors in essential tremor. Brain 135(1):105–116

14. Goetz CG, Tilley BC, Shaftman SR, Stebbins GT, Fahn S, Martinez-Martin P, Poewe W, Sampaio C, Stern MB, Dodel R, Dubois B, Holloway R, Jankovic J, Kulisevsky J, Lang AE, Lees A, Leurgans S, LeWitt PA, Nyenhuis D, Olanow CW, Rascol O, Schrag A, Teresi JA, van Hilten JJ, LaPelle N (2008) Movement Disorder Society-sponsored revision of the Unified Parkinson's Disease Rating Scale (MDS-UPDRS): scale presentation and clinimetric testing results. Mov Disord 23(15):2129–2170
15. Fahn S, Tolosa E, Marin C (1988) Clinical rating scale for tremor. In: Parkinson's disease and movement disorders. Urban & Schwarzenberg, Baltimore, pp 225–234
16. Elble RJ (2016) The essential tremor rating assessment scale. J Neurol Neuromed 1(4):34–38
17. Movement Disorder Society Task Force on Rating Scales for Parkinson's Disease (2003) The Unified Parkinson's Disease Rating Scale (UPDRS): status and recommendations. Mov Disord 18(7):738–750
18. Fahn S, Elton RL, UPDRS Program Members (1987) Unified Parkinson's Disease Rating Scale. In: Recent developments in Parkinson's disease, Vol. 2. Macmillan Healthcare Information, Florham Park, pp 153–163, 293–304.
19. LeMoyne R (2013) Wearable and wireless accelerometer systems for monitoring Parkinson's disease patients—a perspective review. Adv Park Dis 2(4):113–115
20. Giller CA, Dewey RB, Ginsburg MI, Mendelsohn DB, Berk AM (1998) Stereotactic pallidotomy and thalamotomy using individual variations of anatomic landmarks for localization. Neurosurgery 42(1):56–65
21. Niranjan A, Kondziolka D, Baser S, Heyman R, Lunsford LD (2000) Functional outcomes after gamma knife thalamotomy for essential tremor and MS-related tremor. Neurology 55(3):443–446
22. Young RF, Jacques S, Mark R, Kopyov O, Copcutt B, Posewitz A, Li F (2000) Gamma knife thalamotomy for treatment of tremor: long-term results. J Neurosurg 93(S3):128–135
23. Hossen A, Muthuraman M, Al-Hakim Z, Raethjen J, Deuschl G, Heute U (2013) Discrimination of Parkinsonian tremor from essential tremor using statistical signal characterization of the spectrum of accelerometer signal. Biomed Mater Eng 23(6):513–531
24. LeMoyne R, Mastroianni T (2017) Smartphone and portable media device: a novel pathway toward the diagnostic characterization of human movement. In: Smartphones from an applied research perspective. InTech, Rijeka, Croatia, pp 1–24
25. LeMoyne R, Mastroianni T (2017) Wearable and wireless gait analysis platforms: smartphones and portable media devices. In: Wireless MEMS networks and applications. Elsevier, New York, pp 129–152
26. LeMoyne R, Mastroianni T (2016) Telemedicine perspectives for wearable and wireless applications serving the domain of neurorehabilitation and movement disorder treatment. In: Telemedicine SMGroup, Dover, Delaware, pp 1–10
27. LeMoyne R, Mastroianni T (2015) Use of smartphones and portable media devices for quantifying human movement characteristics of gait, tendon reflex response, and Parkinson's disease hand tremor. In: Mobile health technologies, methods and protocols. Springer, New York, pp 335–358
28. LeMoyne R, Coroian C, Cozza M, Opalinski P, Mastroianni T, Grundfest W (2009) The merits of artificial proprioception, with applications in biofeedback gait rehabilitation concepts and movement disorder characterization. In: Biomedical engineering. InTech, Vienna, pp 165–198
29. LeMoyne R, Mastroianni T, Cozza M, Coroian C, Grundfest W (2010) Implementation of an iPhone for characterizing Parkinson's disease tremor through a wireless accelerometer application. In: 32nd Annual international conference of the IEEE, Engineering in Medicine and Biology Society (EMBS), pp 4954–4958
30. LeMoyne R, Mastroianni T, Grundfest W (2013) Wireless accelerometer configuration for monitoring Parkinson's disease hand tremor. Adv Park Dis 2(2):62–67
31. LeMoyne R, Tomycz N, Mastroianni T, McCandless C, Cozza M, Peduto D (2015) Implementation of a smartphone wireless accelerometer platform for establishing deep brain stimulation treatment efficacy of essential tremor with machine learning. In: 37th Annual international conference of the IEEE, Engineering in Medicine and Biology Society (EMBS), pp 6772–6775

32. LeMoyne R, Mastroianni T, Tomycz N, Whiting D, Oh M, McCandless C, Currivan C, Peduto D (2017) Implementation of a multilayer perceptron neural network for classifying deep brain stimulation in 'On' and 'Off' modes through a smartphone representing a wearable and wireless sensor application. In: 47th Society for Neuroscience annual meeting (featured in Hot Topics; top 1% of abstracts)
33. LeMoyne R, Mastroianni T, McCandless C, Currivan C, Whiting D, Tomycz N (2018) Implementation of a smartphone as a wearable and wireless accelerometer and gyroscope platform for ascertaining deep brain stimulation treatment efficacy of Parkinson's disease through machine learning classification. Adv Park Dis 7(2):19–30
34. Hariz GM, Lindberg M, Bergenheim AT (2002) Impact of thalamic deep brain stimulation on disability and health-related quality of life in patients with essential tremor. J Neurol Neurosurg Psychiatry 72(1):47–52
35. Benabid AL, Pollak P, Louveau A, Henry S, de Rougemont J (1987) Combined (thalamotomy and stimulation) stereotactic surgery of the VIM thalamic nucleus for bilateral Parkinson's disease. Appl Neurophysiol 50(1–6):344–346
36. Volkmann J, Moro E, Pahwa R (2006) Basic algorithms for the programming of deep brain stimulation in Parkinson's disease. Mov Disord 21(S14):S284–S289
37. Amon A, Alesch F (2017) Systems for deep brain stimulation: review of technical features. J Neural Transm 124(9):1083–1091
38. Isaias IU, Tagliati M (2008) Deep brain stimulation programming for movement disorders. In: Deep brain stimulation in neurological and psychiatric disorders. Springer, New York, pp 361–397

Chapter 3
Traditional Ordinal Strategies for Establishing the Severity and Status of Movement Disorders, Such as Parkinson's Disease and Essential Tremor

Abstract Ordinal scale strategies are standardly applied to diagnose the severity of neurodegenerative movement disorders, such as Parkinson's disease and Essential tremor. A clinician is tasked with the challenge of assigning an ordinal parameter based on a series of criteria to quantify a subjectively observed interpretation. Multiple ordinal scale systems exist for evaluating movement disorder symptoms. However, the issue is the uncertainty of translating the findings of one scale to another. The Unified Parkinson's Disease Rating Scale (UPDRS) and upgraded Movement Disorder Society Unified Parkinson's Disease Rating Scale (MDS-UPDRS) are commonly utilized for evaluating Parkinson's disease severity. The Fahn-Tolosa-Marin Tremor Rating Scale is prevalently applied for Essential tremor. There are issues of concern regarding the application of ordinal scale approaches for determining the state of progressive neurodegenerative movement disorders, such as Parkinson's disease and Essential tremor. The reliability of ordinal scale systems has not been conclusively established, and interpretive disparity is apparent respective of experience. A novel resolution is the introduction of wearable and wireless inertial sensor systems to objectively quantify movement disorder tremor. The inertial signal (accelerometer and/or gyroscope) can readily record the intrinsic characteristics of tremor for both Parkinson's disease and Essential tremor. Successful testing and evaluation have even demonstrated the efficacy of deep brain stimulation systems for Parkinson's disease and Essential tremor using a smartphone as a wearable and wireless inertial sensor system. These findings enable the pathways for developing Network Centric Therapy, which is in essence the emergence of the Internet of Things for healthcare regarding the domains of robustly diagnosing severity of neurodegenerative movement disorders, such as Parkinson's disease and Essential tremor.

Keywords Ordinal scale · Parkinson's disease · Essential tremor · Unified Parkinson's Disease Rating Scale (UPDRS) · Movement Disorder Society Unified Parkinson's Disease Rating Scale (MDS-UPDRS) · Fahn-Tolosa-Marin Tremor Rating Scale · Wearable and wireless system · Network Centric Therapy

© Springer Nature Singapore Pte Ltd. 2019   25
R. LeMoyne et al., *Wearable and Wireless Systems for Healthcare II*,
Smart Sensors, Measurement and Instrumentation 31,
https://doi.org/10.1007/978-981-13-5808-1_3

3.1   Introduction

The traditional means for establishing the status of a neurodegenerative movement disorder is through the application of an ordinal scale. The clinician applies expert although subjective interpretation of the movement disorder symptoms according to a series of criteria that define the nature of the ordinal scale [1]. Two types of neurodegenerative movement disorder are addressed: Parkinson's disease and Essential tremor. Regarding Parkinson's disease multiple scales exist, for which there is an issue of contrasting and comparing results between these scales [2]. Furthermore, the determination of reliability and variability respective for levels of clinical expertise is a subject of concern [2, 3]. Prevalent ordinal scale techniques for assessing Parkinson's disease status are the Unified Parkinson's Disease Rating Scale (UPDRS) and Movement Disorder Society Unified Parkinson's Disease Rating Scale (MDS-UPDRS) [2, 4, 5]. The Fahn-Tolosa-Marin Tremor Rating Scale is a prevalent means for evaluating Essential tremor severity [6, 7]. A novel alternative approach to quantifying tremor for neurodegenerative movement disorders, such as Parkinson's disease and Essential tremor, is the application of wearable and wireless inertial sensors [8–14]. The presence of wearable and wireless inertial sensors has been further extended toward using the smartphone as a wireless inertial sensor platform for the determination of deep brain stimulation system efficacy through machine learning classification [12–14]. The advent of wearable and wireless systems for the objective and quantified measurement of tremor for neurodegenerative movement disorders, such as Parkinson's disease and Essential tremor, establishes the precedence for the foundation of Network Centric Therapy.

3.2   Clinical Assessment of Parkinson's Disease

The use of an ordinal scale is the widely accepted present strategy for clinically determining the severity of Parkinson's disease symptoms, such as tremor. A clinician subjectively interprets characteristics of the movement disorder in accordance to a series of criteria assigned to an ordinal scale rating. However, this technique only provides a temporal snapshot in light of the prescribed clinical meeting rather than assessment of an inherently fluctuating neurodegenerative disease [1]. Figure 3.1 provides a schematic of the implementation of the ordinal scale process. Multiple ordinal scale strategies are available at the clinician's discretion to provide a historical perspective of the movement disorder for the objective of developing an efficacious treatment strategy [2].

Spanning from 1960 and beyond, 11 rating scales for quantifying Parkinson's disease have been identified. These scales pertain to the assessment of impairment and disability for people with Parkinson's disease symptoms. Ramaker et al. further distinguish these 11 ordinal scales into three types:

- Impairment Scales
- Disability Scales
- Impairment and Disability Sections in Multimodular Scales [2]

Fig. 3.1 Methodology for
the implementation of an
ordinal scale rating for
movement disorder

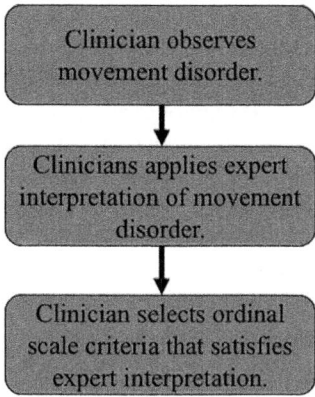

The evaluation of impairment addressed in these three scales are:

- Webster Scale
- Columbia University Rating Scale (CURS)
- Parkinson's Disease Impairment Scale

Regarding the quantification of disability there are four scales:

- Schwab and England Scale
- Northwestern University Disability Scale (NUDS)
- Intermediate Scale for Assessment of Parkinson's Disease
- Extensive Disability Scale

In terms of ascertaining both impairment and disability for Parkinson's disease, there are four available scales:

- New York University Parkinson's Disease Evaluation
- University of California Los Angeles Scale
- Short Parkinson Evaluation Scale
- Unified Parkinson's Disease Rating Scale (UPDRS) [2]

These scales present a wide range of reliability based on published results. With respect to these three formats of evaluating Parkinson's disease, the CURS, NUDS, and UPDRS are the most prevalent with moderate to good reliability and validity. These scales can be challenging to contrast, since they can differ in terms of number of items to be assessed, scoring, and grading [2]. Furthermore, an ordinal assignment to a complex neurodegenerative disease inherently provides minimal insight regarding the efficacy of a prescribed treatment strategy.

Multiple attempts have been conducted to establish the reliability of the ordinal scale approach, such as the UPDRS. However, achieving conclusive levels of consistent reliability appears to be a matter of contention as multiple published studies tend to yield contradictory results [2]. Siderowf et al. conducted a study that attained excellent test-retest reliability from an intra-examiner perspective for patients with early-stage Parkinson's disease symptoms with respect to the activities of daily living, motor, and total UPDRS [15]. Findings of Metman et al. suggest a high level of reliability for the motor aspect of the UPDRS [16]. Other studies, such as by Richards et al., reveal varying levels of reliability throughout certain aspects of the UPDRS motor examination [17].

A further concern is the availability of resources to conduct ordinal scale evaluations. For example, highly experienced movement disorder specialists may be the most sought after, yet these specialists are least available in consideration of the stratification of experience. Post et al. ascertain a notable disparity with scoring interpretation of the UPDRS motor examination with respect to evaluators of varying levels of expertise, such as nurse practitioners, residents in neurology, movement disorder specialists, and senior movement disorder specialists [3]. These findings emphasize the significance of the selection of raters, especially in the context of the longitudinal span of the intrinsically progressive characteristics of a neurodegenerative movement disorder.

The development of the Unified Parkinson's Disease Rating Scale (UPDRS) occurred during the 1980s [18, 19]. The UPDRS has become the prevalent clinical rating scale for the assessment of Parkinson's disease [4]. During 2003, the Movement Disorder Society Task Force for Rating Scales for Parkinson's disease recommended in light of thorough consideration of the strengths and weaknesses of the considerably utilized UPDRS that an evolved version be established. The Movement Disorder Society (MDS) has further evolved the UPDRS to the MDS-UPDRS [5]. The MDS-UPDRS is composed of four segmented aspects as presented in Fig. 3.2.

These four aspects of the MDS-UPDRS each characterize specific features of Parkinson's disease. Part I: non-motor experiences of daily living of the MDS-UPDRS is determined by both the patient/caregiver and investigator. Part II: motor experiences of daily living is envisioned to be defined by the patient/caregiver without supervision from an investigator. Part III: motor examination assesses

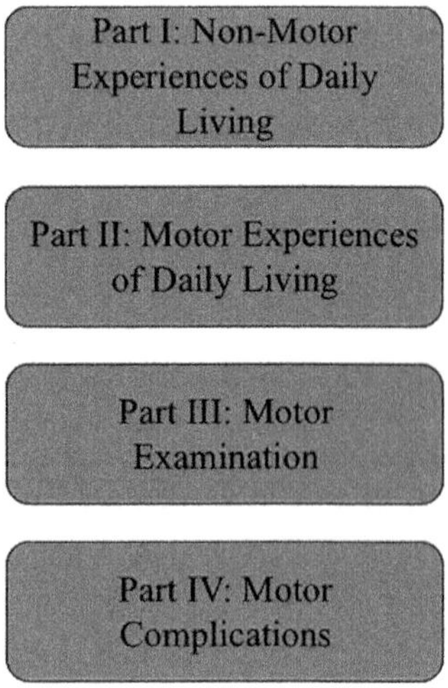

Fig. 3.2 Four segments comprising the MDS-UPDRS [19]

Part I: Non-Motor Experiences of Daily Living

Part II: Motor Experiences of Daily Living

Part III: Motor Examination

Part IV: Motor Complications

movement quality of domains, such as gait and tremor. Part IV: motor complications define fluctuation frequency and duration of "On" and "Off" states [19].

The MDS-UPDRS is envisioned to be a reasonably time efficient means of clinically evaluating Parkinson's disease. Part I (non-motor experiences of daily living) is estimated to take on the order of 10 min. Part III (motor examination) and Part IV (motor complications) require approximately 15 min and 5 min, respectively. In summary, these aspects of the MDS UPDRS are estimated to span roughly a half hour [19].

The MDS-UPDRS is organized to a five-point ordinal scale spanning from 0–4. The ordinal scale ratings are characterized by the following descriptors characterized in Fig. 3.3. The ordinal scale ratings that are indicative of movement disorder symptoms are further clarified. Symptoms that are bounded to low intensity or frequency resulting in no influence on function pertain to "1 (slight)." For the next rating, "2 (mild)" involves symptoms that result in modest influence on function in terms frequency or intensity. In terms of frequency and intensity, "3 (moderate)" and "4 (severe)" are reserved for scenarios for which functionality is considerably impacted and prevented, respectfully. Clinical evaluation supports the validity and reliability of the MDS-UPDRS [19]. Note that an issue with this ordinal scale technique is that it only provides the clarity of five ordinal parameters for describing the considerable complexities of Parkinson's disease.

Martínez-Martín et al. advocate further refinement of the MDS-UPDRS. They recommend the identification of delineating points to identify distinction between mild, moderate, and severe stages of Parkinson's disease. These delineators are proposed for all four segments of the MDS-UPDRS [20]. Researchers also advocate the subjective examination-derived ordinal scale methodology and objective sensor-based approach in tandem, which possess complimentary roles for robust

Fig. 3.3 MDS-UPDRS
five-point ordinal scale
[19]

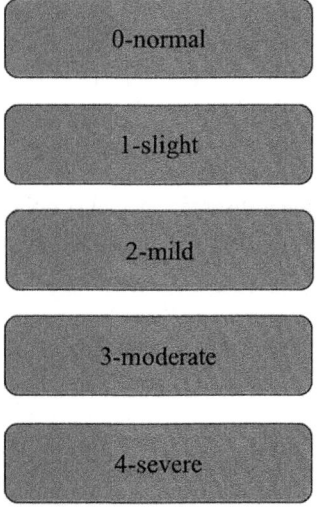

treatment intervention of neurodegenerative movement disorders, such as Parkinson's disease [21].

Other attempts have been made to use inertial sensor systems, such as an accelerometer, to quantify segments of the MDS-UPDRS. One example is the finger-tapping task, which is an aspect of the MDS-UPDRS. Stamatakis et al. successfully demonstrated a computational algorithm for deriving MDS-UPDRS finger-tapping task scores based on the objectively quantified acceleration signal. This perspective offers an inexpensive and simplified approach for the medical community, respective of Parkinson's disease [22].

3.3   Clinical Assessment of Essential Tremor

Essential tremor and Parkinson's disease are both clinically identifiable by disparate tremor symptoms. They are attributed to dysfunction of distinct neuroanatomical structures. Furthermore, Essential tremor and Parkinson's disease are treated by different medical intervention strategies [23]. Intuitively, the application of a unique ordinal scale strategy would likewise be appropriate for the diagnosis of Essential tremor severity.

A widely applied ordinal scale approach for evaluating Essential tremor severity is the Fahn-Tolosa-Marin Tremor Rating Scale [6, 7]. The scale applies a five-point ordinal approach ranging from 0 to 4. There are three primary segments to the scale, such as rest, postural, and kinetic tremor pertaining to specific aspects of the anatomy, tremor influencing motor activities (handwriting, drawing, and pouring water), and daily living activities. The ranking of tremor is bounded by the perceived amplitude of the tremor as described in Fig. 3.4 [24]. The Essential Tremor Rating Assessment Scale (TETRAS) offers more bounding thresholds to provide greater characterization, which is advocated as being more capable for evaluating severe Essential tremor (Fig. 3.5) [25].

A notable issue with the Fahn-Tolosa-Marin Tremor Rating Scale is the inherently subjective nature of interpreting the appropriate ordinal ranking based on the defined criteria. Furthermore, this ordinal strategy only applies five parameters to define a complex neurodegenerative movement disorder, such as Essential tremor. The scale inherently requires a predetermined appointment with preferably a movement disorder specialist rather than a convenient means of autonomously monitoring a subject with a neurodegenerative movement disorder.

3.4   Wearable and Wireless Systems for the Quantification of Movement Disorder Tremor

An alternative to the traditional ordinal scale approach is the wearable and wireless system consisting of inertial sensors to quantify movement, such as tremor. Wearable and wireless systems for quantifying tremor can be readily applied in a homebound

Fig. 3.4 Tremor amplitude thresholds for defining ordinal scale criteria for the original Fahn-Tolosa-Marin Tremor Rating Scale [24]

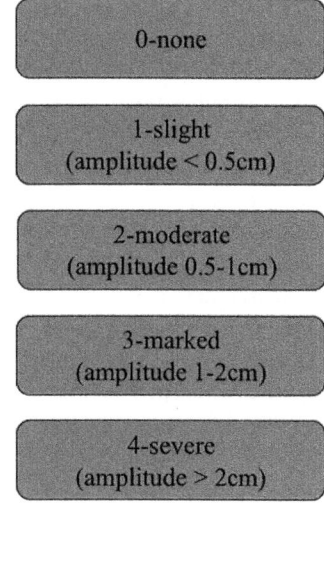

0-none

1-slight
(amplitude < 0.5cm)

2-moderate
(amplitude 0.5-1cm)

3-marked
(amplitude 1-2cm)

4-severe
(amplitude > 2cm)

Fig. 3.5 The Essential Tremor Rating Assessment Scale (TETRAS) amplitude thresholds for the defining ordinal scale criteria regarding tremor of the upper limb [25]

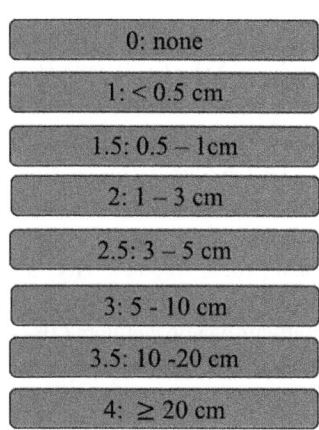

0: none

1: < 0.5 cm

1.5: 0.5 – 1cm

2: 1 – 3 cm

2.5: 3 – 5 cm

3: 5 - 10 cm

3.5: 10 -20 cm

4: ≥ 20 cm

setting. The acquired signal data characterizing the movement disorder can be conveyed by wireless means to a local computer or an Internet resource, such as a Cloud database anywhere in the world, for post-processing. Considerable testing and evaluation have successfully demonstrated the potential of wireless accelerometers and smartphones in particular as wearable and wireless systems for quantifying movement disorder attributes [1, 26–31].

The origins of wearable and wireless systems originated by LeMoyne et al. derive from the objective to provide an alternative to the ordinal scale technique for the quantification of the deep tendon reflex, which can be applied for characterizing spastic movement disorder demonstrated by a hemiplegic reflex pair. The wireless quantified reflex device was progressively evolved over the course of four generations, while successfully quantifying the patellar tendon reflex. The fundamental aspects of the device are an impact pendulum to evoke the patellar tendon reflex and a wearable and wireless accelerometer mounted about the lateral malleolus of the ankle [32–37]. The concept of the wireless quantified reflex device has been extrapolated to other types of wearable and wireless systems, such as smartphones and portable media devices, while even achieving considerable machine learning classification accuracy for distinguishing between a patellar tendon hemiplegic reflex pair [38–42].

With the successful implementation of the wireless quantified reflex device, LeMoyne et al. sought to apply this perspective to the domain of movement disorders. Wearable and wireless systems, such as smartphones and wireless accelerometers, mounted about the dorsum of the hand have been successfully demonstrated for quantifying Parkinson's disease tremor and simulated tremor [8–11]. With the amalgamation of machine learning techniques, the smartphone as a wearable and wireless inertial sensor (accelerometer and/or gyroscope) system has successfully attained considerable machine learning classification accuracy for distinguishing between deep brain stimulation system "On" and "Off" status for both Parkinson's disease and Essential tremor [12–14].

3.5   Extrapolation to Network Centric Therapy

The emergence of wearable and wireless systems, such as the smartphone, establishes the foundation for Network Centric Therapy. Signal data acquired for recording tremor severity for neurodegenerative movement disorders, such as Parkinson's disease and Essential tremor, can be conveyed to Cloud computing databases for post-processing. Not only can this paradigm enable high-fidelity and patient-specific status tracking for robust diagnosis and prognosis, but on a larger scale, many patients can be evaluated from a data science perspective. The opportunities enabled through data science may provide treatment optimization and more novel therapy strategies.

3.6   Conclusion

Traditional medical practice involves applying an ordinal scale technique to quantify the status of neurodegenerative movement disorders, such as Parkinson's disease and Essential tremor. The objective of an expert clinician is to provide an interpretation of the observed movement disorder status according to a series of criteria. The Unified Parkinson's Disease Rating Scale (UPDRS) and upgraded Movement Disorder Society Unified Parkinson's Disease Rating Scale (MDS-UPDRS) are standard for evaluating Parkinson's disease severity. The Fahn-Tolosa-Marin Tremor Rating Scale is predominant for Essential tremor. Although this process has been applied for a considerable span of time, this approach is inherently subjective. There exist multiple ordinal scales, for which the translation between scales is not robustly defined. Furthermore, the ordinal scale strategy restrains the characterization of complex neurodegenerative movement disorders, such as Parkinson's disease and Essential tremor, to merely five ordinal parameters. The reliability of these ordinal scales is a subject of contention, and disparity of interpretation is evident based on the experience of the evaluator.

Wearable and wireless systems using inertial sensors, such as the accelerometer and gyroscope, offer the potential to substantially advance the diagnosis and evaluation of neurodegenerative movement disorders, such as Parkinson's disease and Essential tremor. Recent research endeavors have demonstrated the smartphone in the context of a wearable and wireless system for quantifying Parkinson's disease and Essential tremor. This capability has been extended to the machine learning classification of deep brain stimulation efficacy. A notable observation is the ability to remotely locate the evaluation at a site of the patient's preference with the post-processing resources situated anywhere in the world, which is a logical observation based on the capabilities of Internet connectivity and data storage.

These trends implicate the nascent state of Network Centric Therapy. Network Centric Therapy represents the application of the Internet of Things for healthcare respective of the domains of neurodegenerative movement disorders, such as Parkinson's disease and Essential tremor. Inertial sensor signal data can be wirelessly conveyed to Cloud computing databases through wearable and wireless systems, such as a smartphone. The Cloud database can be historically tracked for patient-specific diagnostic and prognostic analysis. With the collection of many patients, data science may incorporate and elucidate new perspectives for therapy strategies. Patient-specific interventions may be continuously and progressively optimized based on the wealth of signal data provided by wearable and wireless systems for measuring neurodegenerative movement disorders, such as Parkinson's disease and Essential tremor.

References

1. LeMoyne R, Coroian C, Cozza M, Opalinski P, Mastroianni T, Grundfest W (2009) The merits of artificial proprioception, with applications in biofeedback gait rehabilitation concepts and movement disorder characterization. In: Biomedical engineering. InTech, Vienna, pp 165–198
2. Ramaker C, Marinus J, Stiggelbout AM, Van Hilten BJ (2002) Systematic evaluation of rating scales for impairment and disability in Parkinson's disease. Mov Disord 17(5):867–876
3. Post B, Merkus MP, de Bie RM, de Haan RJ, Speelman JD (2005) Unified Parkinson's disease rating scale motor examination: are ratings of nurses, residents in neurology, and movement disorders specialists interchangeable? Mov Disord 20(12):1577–1584
4. Goetz CG, Stebbins GT, Chmura TA, Fahn S, Poewe W, Tanner CM (2010) Teaching program for the Movement Disorder Society-sponsored revision of the Unified Parkinson's Disease Rating Scale: (MDS-UPDRS). Mov Disord 25(9):1190–1194
5. Movement Disorder Society Task Force on Rating Scales for Parkinson's Disease (2003) The Unified Parkinson's Disease Rating Scale (UPDRS): status and recommendations. Mov Disord 18(7):738–750

6. López-Blanco R, Velasco MA, Méndez-Guerrero A, Romero JP, del Castillo MD, Serrano JI, Benito-León J, Bermejo-Pareja F, Rocon E (2018) Essential tremor quantification based on the combined use of a smartphone and a smartwatch: the NetMD study. J Neurosci Methods 303:95–102

7. Zheng X, Vieira Campos A, Ordieres-Meré J, Balseiro J, Labrador Marcos S, Aladro Y (2017) Continuous monitoring of essential tremor using a portable system based on smartwatch. Front Neurol 8(96):1–9

8. LeMoyne R, Coroian C, Mastroianni T (2009) Quantification of Parkinson's disease characteristics using wireless accelerometers. In: ICME International conference on IEEE Complex Medical Engineering (CME), pp 1–5

9. LeMoyne R, Mastroianni T, Cozza M, Coroian C, Grundfest W (2010) Implementation of an iPhone for characterizing Parkinson's disease tremor through a wireless accelerometer application. In: 32nd Annual international conference of the IEEE, Engineering in Medicine and Biology Society (EMBS), pp 4954–4958

10. LeMoyne R (2013) Wearable and wireless accelerometer systems for monitoring Parkinson's disease patients—a perspective review. Adv Park Dis 2(4):113–115

11. LeMoyne R, Mastroianni T, Grundfest W (2013) Wireless accelerometer configuration for monitoring Parkinson's disease hand tremor. Adv Park Dis 2(2):62–67

12. LeMoyne R, Tomycz N, Mastroianni T, McCandless C, Cozza M, Peduto D (2015) Implementation of a smartphone wireless accelerometer platform for establishing deep brain stimulation treatment efficacy of essential tremor with machine learning. In 37th Annual international conference of the IEEE, Engineering in Medicine and Biology Society (EMBS), pp 6772–6775

13. LeMoyne R, Mastroianni T, Tomycz N, Whiting D, Oh M, McCandless C, Currivan C, Peduto D (2017) Implementation of a multilayer perceptron neural network for classifying deep brain stimulation in 'On' and 'Off' modes through a smartphone representing a wearable and wireless sensor application. In: 47th Society for Neuroscience annual meeting (featured in Hot Topics; top 1% of abstracts)

14. LeMoyne R, Mastroianni T, McCandless C, Currivan C, Whiting D, Tomycz N (2018) Implementation of a smartphone as a wearable and wireless accelerometer and gyroscope platform for ascertaining deep brain stimulation treatment efficacy of Parkinson's disease through machine learning classification. Adv Park Dis 7(2):19–30

15. Siderowf A, McDermott M, Kieburtz K, Blindauer K, Plumb S, Shoulson I (2002) Test–retest reliability of the unified Parkinson's disease rating scale in patients with early Parkinson's disease: results from a multicenter clinical trial. Mov Disord 17(4):758–763

16. Metman LV, Myre B, Verwey N, Hassin-Baer S, Arzbaecher J, Sierens D, Bakay R (2004) Test–retest reliability of UPDRS-III, dyskinesia scales, and timed motor tests in patients with advanced Parkinson's disease: an argument against multiple baseline assessments. Mov Disord 19(9):1079–1084

17. Richards M, Marder K, Cote L, Mayeux R (1994) Interrater reliability of the Unified Parkinson's Disease Rating Scale motor examination. Mov Disord 9(1):89–91

18. Fahn S, Elton RL, UPDRS Program Members (1987) Unified Parkinson's Disease Rating Scale. In: Recent developments in Parkinson's disease, Vol. 2. Macmillan Healthcare Information, Florham Park, pp 153–163, 293–304.

19. Goetz CG, Tilley BC, Shaftman SR, Stebbins GT, Fahn S, Martinez-Martin P, Poewe W, Sampaio C, Stern MB, Dodel R, Dubois B, Holloway R, Jankovic J, Kulisevsky J, Lang AE, Lees A, Leurgans S, LeWitt PA, Nyenhuis D, Olanow CW, Rascol O, Schrag A, Teresi JA, van Hilten JJ, LaPelle N (2008) Movement Disorder Society-sponsored revision of the Unified Parkinson's Disease Rating Scale (MDS-UPDRS): scale presentation and clinimetric testing results. Mov Disord 23(15):2129–2170

20. Martínez-Martín P, Rodríguez-Blázquez C, Alvarez M, Arakaki T, Arillo VC, Chaná P, Fernández W, Garretto N, Martínez-Castrillo JC, Rodríguez-Violante M, Serrano-Duenas M,

Ballesteros D, Rojo-Abuin JM, Chaudhuri KR, Merello M (2015) Parkinson's disease severity levels and MDS-Unified Parkinson's Disease Rating Scale. Parkinsonism Relat Disord
21(1):50–54

21. Bhidayasiri R, Martinez-Martin P (2017) Clinical assessments in Parkinson's disease: scales
and monitoring. Int Rev Neurobiol 132:129–182

22. Stamatakis J, Ambroise J, Crémers J, Sharei H, Delvaux V, Macq B, Garraux G (2013) Finger
tapping clinimetric score prediction in Parkinson's disease using low-cost accelerometers.
Comput Intell Neurosci Article ID 717853:1–13

23. Habib-ur-Rehman (2000) Diagnosis and management of tremor. Arch Intern Med
160(16):2438–2444

24. Fahn S, Tolosa E, Marin C (1988) Clinical rating scale for tremor. In: Parkinson's disease and
movement disorders. Urban & Schwarzenberg, Baltimore, pp 225–234

25. Elble RJ (2016) The essential tremor rating assessment scale. J Neurol Neuromed 1(4):34–38

26. LeMoyne R, Mastroianni T (2018) Wearable and wireless systems for healthcare I: gait and
reflex response quantification. Springer, Singapore

27. LeMoyne R, Mastroianni T (2017) Smartphone and portable media device: a novel pathway
toward the diagnostic characterization of human movement. In: Smartphone from an applied
research perspective. InTech, Rijeka, Croatia, pp 1–24

28. LeMoyne R, Mastroianni T (2017) Wearable and wireless gait analysis platforms: smartphones and portable media devices. In: Wireless MEMS networks and applications. Elsevier,
New York, pp 129–152

29. LeMoyne R, Mastroianni T (2016) Telemedicine perspectives for wearable and wireless
applications serving the domain of neurorehabilitation and movement disorder treatment. In:
Telemedicine, SMGroup, Dover, Delaware, pp 1–10

30. LeMoyne R, Mastroianni T (2015) Use of smartphones and portable media devices for quantifying human movement characteristics of gait, tendon reflex response, and Parkinson's disease hand tremor. In: Mobile health technologies, methods and protocols. Springer, New York,
pp 335–358

31. LeMoyne R, Coroian C, Mastroianni T, Grundfest W (2008) Accelerometers for quantification
of gait and movement disorders: a perspective review. J Mech Med Biol 8(2):137–152

32. LeMoyne RC (2010) Wireless quantified reflex device. Ph.D. Dissertation UCLA

33. LeMoyne R, Mastroianni T, Coroian C, Grundfest W (2011) Tendon reflex and strategies for
quantification, with novel methods incorporating wireless accelerometer reflex quantification
devices, a perspective review. J Mech Med Biol 11(3):471–513

34. LeMoyne R, Mastroianni T, Kale H, Luna J, Stewart J, Elliot S, Bryan F, Coroian C, Grundfest
W (2011) Fourth generation wireless reflex quantification system for acquiring tendon reflex
response and latency. J Mech Med Biol 11(1):31–54

35. LeMoyne R, Coroian C, Mastroianni T, Grundfest W (2008) Quantified deep tendon reflex
device for response and latency, third generation. J Mech Med Biol 8(4):491–506

36. LeMoyne R, Dabiri F, Jafari R (2008) Quantified deep tendon reflex device, second generation.
J Mech Med Biol 8(1):75–85

37. LeMoyne R, Jafari R, Jea D (2005) Fully quantified evaluation of myotatic stretch reflex. In:
35th Society for Neuroscience annual meeting

38. LeMoyne R, Mastroianni T (2017) Implementation of a smartphone wireless gyroscope platform with machine learning for classifying disparity of a hemiplegic patellar tendon reflex
pair. J Mech Med Biol 17(6):1750083

39. LeMoyne R, Kerr W, Zanjani K, Mastroianni T (2014) Implementation of an iPod wireless
accelerometer application using machine learning to classify disparity of hemiplegic and
healthy patellar tendon reflex pair. J Med Imaging Health Inform 4(1):21–28

40. LeMoyne R, Mastroianni T (2016) Smartphone wireless gyroscope platform for machine
learning classification of hemiplegic patellar tendon reflex pair disparity through a multilayer
perceptron neural network. In: Wireless Health (WH) of IEEE, pp 103–108

41. LeMoyne R, Mastroianni T (2014) Implementation of a smartphone as a wireless gyroscope application for the quantification of reflex response. In: 36th Annual international conference of the IEEE, Engineering in Medicine and Biology Society (EMBS), pp 3654–3657
42. LeMoyne R, Mastroianni T, Grundfest W, Nishikawa K (2013) Implementation of an iPhone wireless accelerometer application for the quantification of reflex response. In: 35th Annual international conference of the IEEE, Engineering in Medicine and Biology Society (EMBS), pp 4658–4661

Chapter 4
Deep Brain Stimulation for the Treatment of Movement Disorder Regarding Parkinson's Disease and Essential Tremor with Device Characterization

Abstract Deep brain stimulation has provided an efficacious alternative for the treatment of progressive neurodegenerative movement disorders, such as Parkinson's disease and Essential tremor, for more than a quarter of a century. This intervention strategy offers an adjustable and even reversible therapy by contrast to the permanency of lesion inducing neurosurgery, especially if medication has been deemed intractable. Furthermore, deep brain stimulation has been affirmed as a long-term means of suppressing movement disorder symptoms, although the foundational mechanisms remain to be conclusively ascertained. Target selection is specific to the type of neurodegenerative movement disorder diagnosed, and primary risks pertain to surgery and adverse effects resultant from the activation of stimulation. The general aspects that comprise the deep brain stimulation system are discussed from a system-level perspective, such as the implantable pulse generator, battery, connecting wire, and electrode lead. Even after the expert implantation of a deep brain stimulation system, the acquisition of an optimal parameter configuration presents a rather daunting and time-consuming process even for the highly talented and skilled clinician. The sheer quantity of permutations for optimizing parameters, such as polarity, amplitude of stimulation, rate of stimulation, and pulse width, presents a labor-intensive endeavor. The parameter configuration and operation of the deep brain stimulation system are modulated by the clinician programmer and patient programmer. Future concepts for deep brain stimulation are discussed, such as closed-loop acquisition of configuration parameters. Foundational to the perspective of deep brain stimulation optimization is the application of wearable and wireless systems, such as the smartphone, for objectively quantified feedback of movement disorder status. These envisioned technology evolutions advocate the prominence of Network Centric Therapy for the treatment of neurodegenerative movement disorders, such as Parkinson's disease and Essential tremor.

Keywords Deep brain stimulation · Parkinson's disease · Essential tremor · Treatment strategy · Implantable pulse generator · Battery · Connecting wire · Electrode lead · Parameter configuration · Optimization · Polarity · Amplitude of stimulation · Rate of stimulation · Pulse width · Clinician programmer · Patient

© Springer Nature Singapore Pte Ltd. 2019   37
R. LeMoyne et al., *Wearable and Wireless Systems for Healthcare II*,
Smart Sensors, Measurement and Instrumentation 31,
https://doi.org/10.1007/978-981-13-5808-1_4

programmer · Closed-loop optimization · Wearable and wireless systems · Smartphone · Quantification · Network Centric Therapy

4.1   Introduction

Deep brain stimulation has provided an efficacious therapy alternative for neurodegenerative movement disorders, by contrast to more conventional interventions, such as medication and lesioning induced by neurosurgery. This technique has become increasingly prevalent over the span of more than a quarter of a century [1–3]. The long-term feasibility of deep brain stimulation for treating Parkinson's disease and Essential tremor has been corroborated, although the fundamental mechanisms remain a subject of research [4–8]. Inherent advantages of deep brain stimulation pertain to its adjustable and even reversible nature [9]. However, there are associated risks regarding the surgical procedure and adverse effects as a result of the stimulation [7]. Intuitively, the target selection for surgical implantation of the deep brain stimulation is dependent on the type of progressive neurodegenerative movement disorder being treated [9, 10].

From a systems perspective, the deep brain stimulation device is composed of an implantable pulse generator powered by a battery that extends to lead electrodes through connecting wire to generate a predetermined electrical signal [11]. The deep brain stimulation system requires programming at the discretion of an expert clinician, and the acquisition of an optimal parameter configuration represents a critical aspect of the efficacy of deep brain stimulation intervention [11–13]. The parameter configuration tuning and optimization process is envisioned to be greatly facilitated through the application of wearable and wireless systems, such as a smartphone, for the quantification of tremor, for which the experimental location and post-processing resources can be remotely situated [14–21]. Future developmental trends for deep brain stimulation envision the development of closed-loop acquisition of parameter configuration [22, 23]. Furthermore, these evolutionary pathways advocate the prominence of Network Centric Therapy for advancing the treatment of progressive neurodegenerative movement disorders, such as Parkinson's disease and Essential tremor.

4.2   The Foundations of Deep Brain Stimulation

Deep brain stimulation for the treatment of Parkinson's disease was successfully developed by the French neurosurgeon Dr. Alim-Louis Benabid during the 1980s. Dr. Alim-Louis Benabid is an Emeritus Professor of Biophysics for the Université Joseph Fourier which is located in Grenoble, France. During the 1980s, Dr. Benabid observed there were primarily two primary techniques for treating Parkinson's disease: medication through Levodopa, which is also referred to as L-dopa, and through neurosurgery. However, neurosurgery, such as thalamic lesioning, comprised associated complications and side effects. Therefore, Dr. Benabid sought to discover a novel intervention technique [1].

Imperative to the neurosurgery lesioning is the proper targeting of thalamus. Prior to inducing the lesion, the region encompassing the target was investigated. By setting the electrode to physiologically relevant frequencies in the 20–50 Hz range, the target would be refined based on the observed response of the patient. Therefore, Dr. Benabid explored the effect of stimulating throughout a range of frequencies, and at 100 Hz, he observed that the movement disorder tremor became suppressed [1].

During 1987, Dr. Benabid and his associated colleagues demonstrated the efficacy of the deep brain stimulation device. For subjects with bilateral Parkinson's disease, the efficacy of the thalamotomy was compared to the application of deep brain stimulation. The findings determined that deep brain stimulation relatively ameliorated tremor symptoms. They applied deep brain stimulation to the ventral intermediate nucleus (VIM) of the thalamus contrasted to VIM thalamotomy, for which the deep brain stimulation application ameliorated tremor. These discoveries represent the seminal methodology of deep brain stimulation that pioneered its introduction to the neuroscience and biomedical community [2, 3].

Later in the 1990s, Dr. Benabid began to investigate a new deep brain stimulation target for Parkinson's disease intervention: the subthalamic nucleus. Dr. Benabid has further contributed to the research and development of deep brain stimulation, such as novel endoventricular entry route, which benefits the implementation procedure of the electrode. Furthermore, as of the timeframe of 2010, Dr. Benabid has considered new neuroanatomical targets for the therapeutic treatment of dystonia and epilepsy [1].

4.3   Long-Term Efficacy and the Quest to Define the Mechanisms of Deep Brain Stimulation

The long-term efficacy of deep brain stimulation for the treatment of neurodegenerative movement disorders, such as Parkinson's disease and Essential tremor, is also a topic that warrants investigation. Research endeavors have ascertained that deep brain stimulation can suppress movement disorder symptoms for Parkinson's disease and Essential tremor with regard to a long-term time perspective [4–7]. Furthermore, the deep brain stimulation configuration parameters were noted to be stable with respect to the protracted longitudinal timeframe [4–6]. However, the fundamental mechanisms for the efficacy of deep brain stimulation to treat progressive neurodegenerative movement disorders have not been conclusively determined. Benabid et al. propose an assortment of plausible mechanisms, such as feedback circuitry disruption [8].

4.4   Benefit and Risk Consideration of Deep Brain Stimulation

Deep brain stimulation for the treatment of movement disorders offers an attractive alternative to conventional ablation techniques, since it is adjustable as implicated by the considerable assortment of parameter configurations. The deep brain stimulation approach is also reversible, because the apparatus can be removed in the event that the intervention strategy is not satisfactory [9]. The intervention through deep brain stimulation has produced notably perceptible advances for quality of life for movement disorders, such as Parkinson's disease [24].

Adverse effects can be delineated into two distinct domains: surgical and stimulation-based adverse effects. Examples of surgical adverse effects can involve scenarios, such as intracerebral hemorrhage and hematoma proximal to the implantable pulse generator. Stimulation-derived adverse effects are generally resolved through the alteration of the parameter configuration. Adverse effects regarding stimulation pertain to:

- Paresthesia (tingling, tickling, pricking sensation)
- Headache
- Dysarthria (characterized by impaired speech)
- Disequilibrium
- Visual disturbances [7]

4.5   Target Selection for Deep Brain Stimulation

4.5.1   Essential Tremor

The target selection of an aspect of the deep brain neuroanatomy is contextually specific to the type of movement disorder. Parkinson's disease and Essential tremor can be treated through the selection of disparate deep brain structures [9, 10]. Essential tremor is primarily treated through stimulating the ventral intermediate nucleus (VIM). The VIM is a motor related aspect of the thalamus, and it receives input from the cerebellum [9, 10]. Projections of the VIM convey to the motor cortex, for which this target is prevalent for the amelioration of Essential tremor [9, 25].

4.5.2   Parkinson's Disease

By contrast, Parkinson's disease has multiple deep brain stimulation targets, such as the ventral intermediate nucleus (VIM) and subthalamic nucleus (STN). Deep brain stimulation treatment using the ventral intermediate nucleus for people with

Parkinson's disease is essentially transitory regarding the amelioration of other associated Parkinson's disease symptoms, such as dyskinesia resulting from drug therapy, rigidity, and bradykinesia. Therefore, the application of deep brain stimulation that targets the VIM for Parkinson's disease is generally reserved to patients, such that their symptoms predominantly emphasize tremor [9].

Further neuroanatomical-targeting alternatives have been the subject of scientific investigation. Two additional targets are the subthalamic nucleus (STN) and globus pallidus internus (GPi). During 1993, Pollak et al. established the feasibility of targeting the subthalamic nucleus for the treatment of Parkinson's disease movement disorder [2, 26]. The subthalamic nucleus is compact in nature, and it is situated about the superior aspect of the midbrain. Furthermore, the subthalamic nucleus has been demonstrated to be influential with regard to the cortical-basal ganglia-thalamocortical pathways [9, 25, 27]. In light of the characteristics of these neurological pathways, clinical efficacy regarding deep brain stimulation applied to the STN has been broadly established [9]. Therapeutic efficacy has been further demonstrated from a long-term perspective with reduction in necessary dopaminergic medical intervention. Also, prolonged amelioration of rigidity and tremor symptoms has been observed [6, 9, 28]. The following year of 1994, the globus pallidus internus was demonstrated by Siegfried and Lippitz as an effective deep brain stimulation target [2, 29].

4.6   Deep Brain Stimulation System Device Description

Deep brain stimulation is primarily applied to ameliorate movement disorder symptoms. As of 2017, deep brain stimulation devices have been approved for treating Parkinson's disease, Essential tremor, dystonia, and epilepsy. Obsessive-compulsive disorder constitutes another approved treatment domain [11].

The standard deep brain stimulation system is composed of multiple aspects. The electrode lead is surgically positioned in a deep neurological structure of the brain. The electrodes are continued by connecting wire to the implantable pulse generator. The implantable pulse generator encases the battery, which represents the energy source, and the electronic circuitry generates the electrical signal that ameliorates movement disorder symptoms [11]. Samples of implantable pulse generator are presented in Fig. 4.1.

Currently, there are three predominant medical device manufacturers for implantable deep brain stimulation systems: Boston Scientific, Medtronic, and St. Jude. Table 4.1 summarizes the companies and their product names. These devices possess similar attributes from a fundamental perspective. However, each device also retains unique technical features. Figure 4.2 illustrates the intrinsic subsystems and features that compose the general deep brain stimulation device [11].

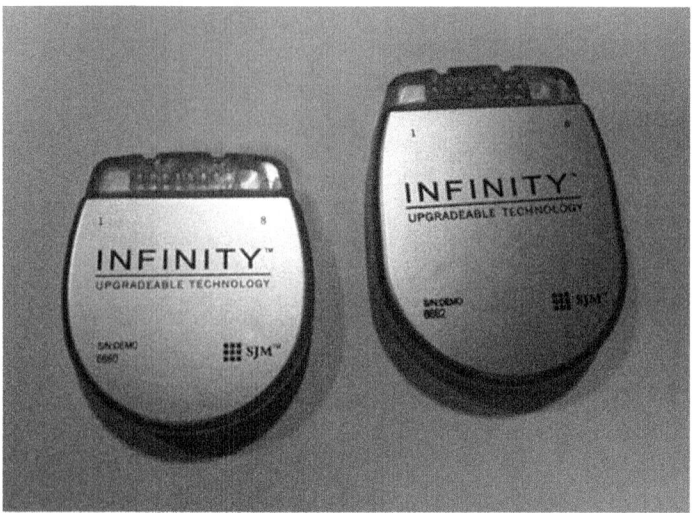

Fig. 4.1 Sample deep brain stimulation implantable pulse generators provided by Allegheny General Hospital

Table 4.1 Medical device manufacturers for implantable deep brain stimulation systems [11]

Company	Name of deep brain stimulation device
Boston Scientific	Vercise
	Vercise PC
	Vercise Gevia
Medtronic	Activa PC
	Activa RC
	Activa SC
St. Jude	Libra
	Libra XP Brio
	Infinity

4.6.1   Electrode Leads and the Implantable Pulse Generator

The electrical characteristics of the electrode leads and the implantable pulse generator define the broader design capabilities of the specific deep brain stimulation system. The geometric design of the electrode constitutes a driving influence for the distribution of the electric signal. Current electrode configurations utilize segmented rings that apply four or eight contacts. The electrode is further characterized in terms of contact length and inter-contact spacing. In summary, the electrode attributes are defined by the number of contacts, contact length, and inter-contact spacing. These characteristics represent significant design features for the deep brain stimulation system, since they cannot be altered once surgically implanted [11]. Figure 4.3 presents an electrode for a deep brain stimulation system.

Fig. 4.2 General perspective of subsystems and features representing a deep brain stimulation device [11]

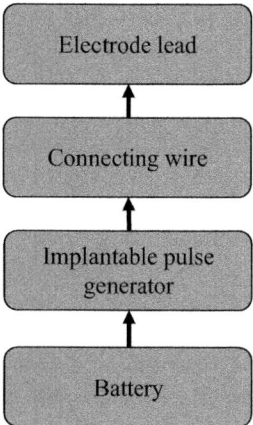

Fig. 4.3 Electrode for a deep brain stimulation system provided by Allegheny General Hospital

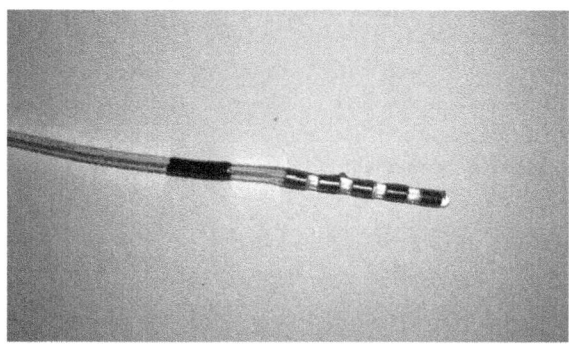

The electrodes in conjunction with the implantable pulse generator can offer either unilateral or bilateral stimulation. Unilateral stimulation pertains to a singular hemisphere of the brain, and bilateral stimulation involves both hemispheres of the brain. Bilateral stimulation applies two sets of electrode leads with connecting wires supported by a singular implantable pulse generator [11]. In the event that bilateral stimulation of the brain's two hemispheres is warranted, a singular implantable pulse generator that issues synchronized commands for two sets of electrode leads would facilitate a simplified strategy as opposed to two implantable pulses generators transmitting signals to two separate unilateral electrode leads.

The polarity of the contacts determines the current distribution about the targeted neural structure. In essence, by influencing the current distribution, the electrode contacts determine the morphology of stimulated neural tissue [11, 30]. There are three configurations regarding polarity: unipolar, bipolar, and multipolar. Polarity is determined as a function of the status of the electrode contacts and implantable pulse generator case, which may be programmed to anode (+), cathode (−), or off (neutral) status [11].

The program configuration can achieve unipolar, bipolar, and multipolar capability. In order to attain a unipolar configuration, the implantable pulse generator functions as the anode, and one electrode contact functions as the cathode. The bipolar configuration emphasizes two electrode contacts with one electrode contact programmed as the anode and the other electrode contact set to the cathode. With regard to the case of the implantable pulse generator, it is assigned to a neutral status. For the multipolar scenario, one of more electrode contacts is set to both anode and cathode status with the current transmitting between the respective cathodes and anodes. An example of a multipolar configuration would be programming two electrode contacts as anodes and another electrode contract as a cathode positioned between the two electrode anode contacts [11].

The source of the electric signal for the deep brain stimulation system derives from the implantable pulse generator. Two contingencies are voltage sources and current source. The resistance is based on intrinsic properties of the neural structures and the interface with the associated electrodes. Therefore, for a voltage source configuration, the intensity of therapeutic current is proportional to the resistance that may vary as a function of time [11, 31, 32]. In light of this potential temporal variation of therapeutic current from a deep brain stimulation system through the use of a voltage source contingency, the current source approach is becoming more frequently applied. The application of a current source approach mitigates the influence of resistance variation by directly regulating current [11, 33].

4.6.2 Battery

The choice of battery type for a deep brain stimulation system represents a significant design consideration. The battery can be either non-rechargeable or rechargeable. The non-rechargeable configuration implicates a design restriction that the limited energy stored can power for the deep brain stimulation system throughout the lifetime of the device. The deep brain stimulation system that applies a rechargeable battery approach offers operational flexibility as the deep brain stimulation device can be periodically recharged for expanded operational lifetime [11].

Battery longevity is intrinsic to multiple aspects of the deep brain stimulation system. The attributes of the stimulating electric signal are influenced by the parameter configuration, such as the polarity, simulation rate, the amplitude of stimulation, and the pulse width. The characteristics of the deep brain stimulation system electronic circuitry impact energy usage. Also, the impedance about the electrodes represents another subject for the overall energy budget that defines system-level requirements for the battery of the deep brain stimulation system, especially with regard to its implementation [11].

4.6.3   Electrical Signal

The electric stimulation field that targets a localized aspect of the brain is derived from the stimulation configuration of the deep brain stimulation system. The four primary configuration parameters are the polarity, amplitude of stimulation, rate of stimulation, and pulse width. The amplitude of stimulation is parameterized in accordance to the source for the electric signal, such as voltage source or current source [11].

With respect to the voltage source approach, the amplitude of stimulation is based on voltage (V). In terms of the current source alternative, the simulation amplitude is provided by milliamperes (mA). The rate of stimulation is defined as the number of pulses per second (Hz). The pulse width is established as the temporal length of the stimulating pulse, which is generally on the scale of microseconds [11].

A stimulation configuration is composed of the parameters defining the electric signal and the polarity setting. A series of stimulation configurations can be sequentially applied in an interleaved manner. Multiple stimulation configurations can be programmed into to a deep brain stimulation system [11].

As an example of the interleaving capability, a deep brain stimulation system could apply two stimulation configurations. These stimulation configurations could be composed of distinct amplitude, pulse rate, and pulse width, such that the stimulation alternates from the first stimulation configuration to the second stimulation configuration. This capability enables the deep brain stimulation system to achieve current distribution suitable to the patient-specific requirements [11].

4.7   Deep Brain Stimulation System Programmer

Communication to the implantable pulse generator of the deep brain stimulation system is achieved through wireless transmission from a programmer [11]. Clinical intervention to modify the stimulation configuration is achieved through a programmer, for which the acquisition of an optimal parameter configuration is an essential task for the application of deep brain stimulation [12, 13]. These programmers are designed for use-control by supervising clinicians and patients undergoing deep brain stimulation therapy. Other versions of deep brain stimulation programmers incorporate Apple digital devices that achieve wireless communication through Bluetooth [11]. The application of Bluetooth wireless connectivity between biomedical devices has been also advocated by LeMoyne and Mastroianni and further demonstrated for the machine learning classification of healthy and neurodegenerative disorder influenced gait with wearable and wireless inertial sensors wirelessly connected by Bluetooth to a portable tablet [34–36]. Figure 4.4 illustrates a representative deep brain stimulation patient programmer, and Fig. 4.5 provides a representative deep brain stimulation clinician programmer.

Fig. 4.4 Representative
deep brain stimulation
patient programmer for
initiation of a
predetermined stimulation
configuration provided by
Allegheny General
Hospital

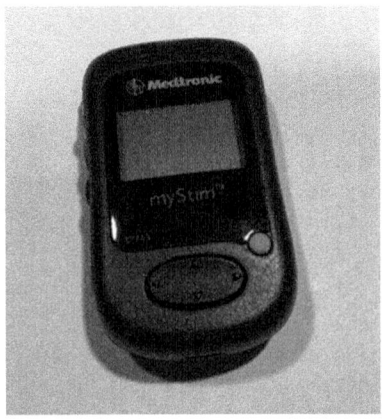

Fig. 4.5 Representative
deep brain stimulation
clinician programmer for
the establishment of a
stimulation parameter
configuration provided by
Allegheny General
Hospital

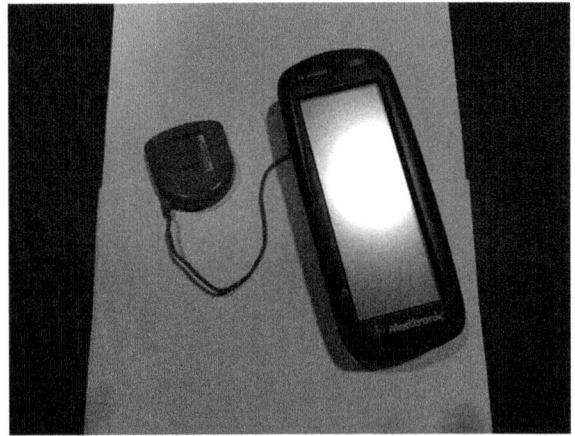

4.8   Attaining the Optimal Parameter Configuration for Deep Brain Stimulation an Issue with Tuning

The current strategy for programming a deep brain stimulation system is inherently iterative. During the preliminary adjustment timeframe, this approach generally requires the patient to frequently interact with clinical resources. The optimization process for ascertaining a suitable parameter configuration can take on the order of several weeks to several months [37]. Panisset et al. have even proposed the concept of assembling deep brain stimulation teams comprised of an array of multidisciplinary specialists for performing tasks, such as the programming phase [38]. Even in light of considerable efficacy of deep brain stimulation for the treatment of movement disorders, the time and resource allocation imperative for managing deep brain stimulation are perceived as relatively substantial. This observation establishes the motivation to simplify the deep brain stimulation parameter optimization process [39].

The implication of the current strategy is the time- and resource-intensive nature of achieving an optimal parameter configuration. Also, there exists a logistical characteristic to the optimization process, such as the distance between a patient and the available clinical resources. The introduction of wearable and wireless systems with wireless Internet connectivity may facilitate the process of ascertaining optimal deep brain stimulation parameter configurations.

The integration of wearable and wireless systems has been demonstrated regarding the assessment of neurodegenerative movement disorders, such as Parkinson's disease and Essential tremor. In particular, the patient's experimental site and post-processing resources can be remotely situated through wireless connectivity to the Internet, such as through an email resource functioning as a provisional Cloud computing resource. The inertial sensor signal data provided by the wearable and wireless system, such as a smartphone, can be transmitted as an email attachment through wireless connectivity to the Internet [14–21]. Machine learning algorithms can provide considerable classification accuracy to distinguish between perceptibly disparate states, such as deep brain stimulation "On" and "Off" status [15–21]. From the perspective of Network Centric Therapy, deep brain stimulation parameter configuration optimization can occur from the wearable and wireless system or Cloud computing environment with the augmented acuity of machine learning.

4.9 Future Perspectives for Deep Brain Stimulation

The perspective for deep brain stimulation of Dr. Hariz spans a commanding quarter of a century. Dr. Hariz envisions the evolution of deep brain stimulation to advanced concepts, such as "on-demand" and closed-loop acquisition of parameter configurations [22]. The current open-loop deep brain stimulation process involves the selection of a parameter configuration that remains static in light of inherent neurological variability. Similar to how the modern pacemaker adapts stimulation variables according to cardiac conditions, the closed-loop deep brain stimulation system would be adaptive to fluctuating neurological conditions [23]. Preliminary research, development, testing, and evaluation imply that the closed-loop strategy transcends the utility of the open-loop approach for deep brain stimulation for the treatment intervention of progressive neurodegenerative movement disorders [40, 41]. These capabilities are in alignment with Network Centric Therapy advocated throughout the subject matter of the book, especially with the amalgamation of deep brain stimulation using wearable and wireless systems as feedback with machine learning classification to distinguish between disparate scenarios.

4.10   Conclusion

Deep brain stimulation has enabled a highly effective and relatively novel treatment strategy for the amelioration of movement disorder symptoms for Parkinson's disease and Essential tremor. The innovative origins of deep brain stimulation can be primarily attributed to the endeavors of Dr. Alim-Louis Benabid, which occurred slightly beyond the span of a quarter of a century ago. This technique has been established as an efficacious therapy for the long-term time perspective.

Deep brain stimulation offers notable utilities that transcend traditional therapy approaches for treating progressive neurodegenerative movement disorders, such as Parkinson's disease and Essential tremor. This novel therapy is pertinent to conditions for which medical intervention is deemed intractable. Furthermore, deep brain stimulation is not only adjustable, but also the device can be removed. These features of deep brain stimulation are stark contrast to the permanency of lesioning neurosurgery.

Since the application of a deep brain stimulation system requires advanced surgical techniques, such as neurosurgery, there are inherent risks that should be addressed relative to the benefits of ameliorating movement disorder symptoms. Generally, primary risks pertain to the actual surgery, such as intracerebral hemorrhaging and the development of a hematoma local to the implantable pulse generator. Other risks involve adverse effects as a consequence of stimulation provided. These surgical techniques apply targeting of an aspect of the deep brain neuroanatomy that is specific to the movement disorder under consideration, such as Parkinson's disease or Essential tremor. The actual mechanisms that associate deep brain stimulation to movement disorder suppression remain to be conclusively defined.

The deep brain stimulation device is further described from a systems approach. A perspective for aspects that comprise the deep brain stimulation system, such as implantable pulse generator, battery, connecting wire, and electrode lead, has been presented. Even with the expert surgical installation of the deep brain stimulation system with precise targeting of the neurological structures, the acquisition of a patient-specific optimal parameter configuration constitutes a laborious and daunting endeavor even for an expert clinician. Current optimization techniques require iterative interaction between patient and clinical resources, which can require on the order of several weeks to several months. A considerable array of configuration parameters exists through the modification of polarity, amplitude of stimulation, rate of stimulation, and pulse width. A means for remote interaction of patient and clinical resources, for which the optimization of the parameter configuration is automated, would be highly desirable.

Recent technology developments advocate the application of wearable and wireless systems, such as the smartphone, for objectively quantifying movement disorder tremor status. This capability establishes the evolution of Network Centric Therapy that involves the Cloud computing storage of inertial sensor signal data that is provided by a wearable and wireless system, such as a smartphone. This capability enables the experimental site and post-processing resources to be remotely situated anywhere in the world. Furthermore, machine learning classification can facilitate the optimization of the deep brain stimulation parameter configuration. These technology advances imply the future development of closed-loop optimization of deep brain stimulation that is uniquely adapted to the specific patient.

References

1. Williams R (2010) Alim-Louis Benabid: stimulation and serendipity. Lancet Neurol 9(12):1152
2. Miocinovic S, Somayajula S, Chitnis S, Vitek JL (2013) History, applications, and mechanisms of deep brain stimulation. JAMA Neurol 70(2):163–171
3. Benabid AL, Pollak P, Louveau A, Henry S, de Rougemont J (1987) Combined (thalamotomy and stimulation) stereotactic surgery of the VIM thalamic nucleus for bilateral Parkinson's disease. Appl Neurophysiol 50(1–6):344–346
4. Rehncrona S, Johnels B, Widner H, Törnqvist AL, Hariz M, Sydow O (2003) Long-term efficacy of thalamic deep brain stimulation for tremor: double-blind assessments. Mov Disord 18(2):163–170
5. Sydow O, Thobois S, Alesch F, Speelman JD (2003) Multicentre European study of thalamic stimulation in essential tremor: a six year follow up. J Neurol Neurosurg Psychiatry 74(10):1387–1391
6. Krack P, Batir A, Van Blercom N, Chabardes S, Fraix V, Ardouin C, Koudsie A, Limousin PD, Benazzouz A, LeBas JF, Benabid AL, Pollak P (2003) Five-year follow-up of bilateral stimulation of the subthalamic nucleus in advanced Parkinson's disease. N Engl J Med 349(20):1925–1934
7. Lyons KE, Koller WC, Wilkinson SB, Pahwa R (2001) Long term safety and efficacy of unilateral deep brain stimulation of the thalamus for parkinsonian tremor. J Neurol Neurosurg Psychiatry 71(5):682–684
8. Benabid AL, Benazzous A, Pollak P (2002) Mechanisms of deep brain stimulation. Mov Disord 17(S3):S73–S74
9. Yu H, Neimat JS (2008) The treatment of movement disorders by deep brain stimulation. Neurotherapeutics 5(1):26–36
10. Hassler R (1959) Anatomy of the thalamus. In: Introduction to stereotaxis with an atlas of the human brain. Thieme, Stuttgart, pp 230–290
11. Amon A, Alesch F (2017) Systems for deep brain stimulation: review of technical features. J Neural Transm 124(9):1083–1091
12. Volkmann J, Moro E, Pahwa R (2006) Basic algorithms for the programming of deep brain stimulation in Parkinson's disease. Mov Disord 21(S14):S284–S289
13. Isaias IU, Tagliati M (2008) Deep brain stimulation programming for movement disorders. In: Deep brain stimulation in neurological and psychiatric disorders. Springer, New York, pp 361–397
14. LeMoyne R, Mastroianni T, Cozza M, Coroian C, Grundfest W (2010) Implementation of an iPhone for characterizing Parkinson's disease tremor through a wireless accelerometer application. In: 32nd Annual international conference of the IEEE, Engineering in Medicine and Biology Society (EMBS), pp 4954–4958
15. LeMoyne R, Tomycz N, Mastroianni T, McCandless C, Cozza M, Peduto D (2015) Implementation of a smartphone wireless accelerometer platform for establishing deep brain stimulation treatment efficacy of essential tremor with machine learning. In: 37th Annual international conference of the IEEE, Engineering in Medicine and Biology Society (EMBS), pp 6772–6775

16. LeMoyne R, Mastroianni T, Tomycz N, Whiting D, Oh M, McCandless C, Currivan C, Peduto D (2017) Implementation of a multilayer perceptron neural network for classifying deep brain stimulation in 'On' and 'Off' modes through a smartphone representing a wearable and wireless sensor application. In: 47th Society for Neuroscience annual meeting (featured in Hot Topics; top 1% of abstracts)

17. LeMoyne R, Mastroianni T, McCandless C, Currivan C, Whiting D, Tomycz N (2018) Implementation of a smartphone as a wearable and wireless accelerometer and gyroscope platform for ascertaining deep brain stimulation treatment efficacy of Parkinson's disease through machine learning classification. Adv Park Dis 7(2):19–30

18. LeMoyne R, Mastroianni T (2017) Smartphone and portable media device: a novel pathway toward the diagnostic characterization of human movement. In: Smartphones from an applied research perspective. InTech, Rijeka, Croatia, pp 1–24

19. LeMoyne R, Mastroianni T (2017) Wearable and wireless gait analysis platforms: smartphones and portable media devices. In: Wireless MEMS networks and applications. Elsevier, New York, pp 129–152

20. LeMoyne R, Mastroianni T (2016) Telemedicine perspectives for wearable and wireless applications serving the domain of neurorehabilitation and movement disorder treatment. In: Telemedicine, SMGroup, Dover, Delaware, pp 1–10

21. LeMoyne R, Mastroianni T (2015) Use of smartphones and portable media devices for quantifying human movement characteristics of gait, tendon reflex response, and Parkinson's disease hand tremor. In: Mobile health technologies, methods and protocols. Springer, New York, pp 335–358

22. Hariz M (2017) My 25 stimulating years with DBS in Parkinson's disease. J Park Dis 7(s1):S33–S41

23. Fang JY, Tolleson C (2017) The role of deep brain stimulation in Parkinson's disease: an overview and update on new developments. Neuropsychiatr Dis Treat 13:723–732

24. Benabid AL (2003) Deep brain stimulation for Parkinson's disease. Curr Opin Neurobiol 13(6):696–706

25. Starr PA, Vitek JL, Bakay RA (1998) Deep brain stimulation for movement disorders. Neurosurg Clin N Am 9(2):381–402

26. Pollak P, Benabid AL, Gross CH, Gao DM, Laurent A, Benazzouz A, Hoffmann D, Gentil M, Perret J (1993) Effets de la stimulation du noyau sous-thalamique dans la maladie de Parkinson. Rev Neurol (Paris) 149(3):175–176

27. Kopell BH, Rezai AR, Chang JW, Vitek JL (2006) Anatomy and physiology of the basal ganglia: implications for deep brain stimulation for Parkinson's disease. Mov Disord 21(S14):S238–S246

28. Schüpbach W, Chastan N, Welter M, Houeto J, Mesnage V, Bonnet A, Czernecki V, Maltête D, Hartmann A, Mallet L, Pidoux B, Dormont D, Navarro S, Cornu P, Mallet A, Agid Y (2005) Stimulation of the subthalamic nucleus in Parkinson's disease: a 5 year follow up. J Neurol Neurosurg Psychiatry 76(12):1640–1644

29. Siegfried J, Lippitz B (1994) Bilateral chronic electrostimulation of ventroposterolateral pallidum: a new therapeutic approach for alleviating all parkinsonian symptoms. Neurosurgery 35(6):1126–1130

30. Butson CR, McIntyre CC (2008) Current steering to control the volume of tissue activated during deep brain stimulation. Brain Stimul 1(1):7–15

31. Benabid AL, Pollak P, Gao D, Hoffmann D, Limousin P, Gay E, Payen I, Benazzouz A (1996) Chronic electrical stimulation of the ventralis intermedius nucleus of the thalamus as a treatment of movement disorders. J Neurosurg 84(2):203–214

32. Hartmann CJ, Wojtecki L, Vesper J, Volkmann J, Groiss SJ, Schnitzler A, Sudmeyer M (2015) Long-term evaluation of impedance levels and clinical development in subthalamic deep brain stimulation for Parkinson's disease. Parkinsonism Relat Disord 21(10):1247–1250

33. Bronstein JM, Tagliati M, McIntyre C, Chen R, Cheung T, Hargreaves EL, Israel Z, Moffitt M, Montgomery EB, Stypulkowski P, Shils J, Denison T, Vitek J, Volkman J, Wertheimer J, Okun

MS (2015) The rationale driving the evolution of deep brain stimulation to constant-current devices. Neuromodulation 18(2):85–89

34. LeMoyne R, Heerinckx F, Aranca T, De Jager R, Zesiewicz T, Saal HJ (2016) Wearable body and wireless inertial sensors for machine learning classification of gait for people with Friedreich's ataxia. In: IEEE 13th International conference on wearable and implantable Body Sensor Networks (BSN), pp 147–151

35. LeMoyne R, Mastroianni T (2018) Bluetooth inertial sensors for gait and reflex response quantification with perspectives regarding cloud computing and the Internet of Things. In: Wearable and wireless systems for healthcare I: gait and reflex response quantification. Springer, Singapore, pp 95–103

36. LeMoyne R, Mastroianni T (2018) Wearable and wireless systems for healthcare I: gait and reflex response quantification. Springer, Singapore

37. Pretto T (2007) Deep brain stimulation. Neurologist 13(2):103–104

38. Panisset M, Picillo M, Jodoin N, Poon YY, Valencia-Mizrachi A, Fasano A, Munhoz R, Honey CR (2017) Establishing a standard of care for deep brain stimulation centers in Canada. Can J Neurol Sci 44(2):132–138

39. Schwalb JM, Hamani C (2008) The history and future of deep brain stimulation. Neurotherapeutics 5(1):3–13

40. Sun FT, Morrell MJ (2014) Closed-loop neurostimulation: the clinical experience. Neurotherapeutics 11(3):553–563

41. Priori A, Foffani G, Rossi L, Marceglia S (2013) Adaptive deep brain stimulation (aDBS) controlled by local field potential oscillations. Exp Neurol 245:77–86

Chapter 5
Surgical Procedure for Deep Brain Stimulation Implantation and Operative Phase with Postoperative Risks

Abstract The surgical procedure for instilling a deep brain stimulation system is an incredibly serious endeavor. A multiphase approach is applied for the implantation of the deep brain stimulation system, such as neurosurgery to position the electrodes and other surgical techniques to implant other aspects of the system. The quality of the surgical procedure can ensure against complication risks, such as infection and hemorrhaging. Electromagnetic interaction can pose hazards to the patient. However, the benefit of noninvasive imaging through magnetic resonance imaging (MRI) transcends the risk in light of the proper safety procedures. Other considerations involve the neurological and neuropsychological effects during the operation of the deep brain stimulation system. By addressing these concerns, a more comprehensive risk to benefit perspective can be established. Finally, a surgical procedure instilled at an internationally renowned hospital is presented. The actual parameter configuration tuning process advocated by the internationally renowned hospital is further discussed.

Keywords Surgical implantation · Deep brain stimulation · Electrode · Implantable pulse generator · Globus pallidus internal segment · Subthalamic nucleus · Surgical risk · Energy interaction · Neurological effects · Neuropsychological effects

5.1 Introduction

The application of a deep brain stimulation is intuitively an extremely serious endeavor. Structures of the deep brain are accessed by electrodes through carefully planned neurosurgery, and the implementation procedure is conducted in a segmented surgical process [1, 2]. Inherent risks associated with the quality of the surgical process are infection and intracranial hemorrhaging [1, 3]. Even after effective surgical implantation, the deep brain stimulation system can potentially adversely interact with energy from extrinsic electromagnetic fields [4, 5]. This observation is in particular a subject of concern for safely applying noninvasive magnetic resonance imaging (MRI) to determine the efficacy of electrode position-ing [6, 7]. Furthermore, there are neurological and neuropsychological effects asso-ciated with the operation of the deep brain stimulation system [1, 8]. With the complexities and risks of implanting and successfully operating a deep brain system presented, the benefit to risk evaluation can be more thoroughly elucidated.

5.2   General Surgical Perspective and Considerations for the Application of Deep Brain Stimulation

The general surgical procedure involves the implantation of the deep brain stimula-tion lead in either the globus pallidus internal segment or the subthalamic nucleus. The lead is extended through the skull by a burr hole and anchored with a fixation device. The lead is attached to a connecting wire. The connecting wire is subcutane-ously conveyed below the scalp through the neck and finally to the anterior aspect of the chest for connection to the implantable pulse generator situated in a surgically developed pocket. With exception to scenarios involving infection, deep brain stim-ulation hardware is infrequently removed [1].

The surgical procedure for a deep brain stimulation system is segmented into primarily two aspects: first, the electrode and then, second, the implantable pulse generator arrangement. The insertion of the electrodes is facilitated by noninvasive stereotactic guidance, such as through magnetic resonance imaging or computed tomography scanning. The two available techniques for implanting the electrodes are through straight or curvilinear incision [2].

Subsequent to the electrode procedure is the insertion of the implantable pulse generator, which can be conducted immediately after the first step of the electrode surgery or externalized over the course of a few days. The insertion of the implant-able pulse generator is achieved through a straight incision. Another step involves joining the electrode lead to implantable pulse generator extension through a subcu-taneous tunnel routed from the scalp to the chest [2].

Implications of the general surgical procedure are the imperative need for advanced imaging techniques for preoperative planning. The application of stereo-tactic systems for neurosurgery is instrumental for the extremely high degree of precision required. Therefore, evolutions for the domain of noninvasive imaging, such as MRI, and stereotactic surgery technologies have the capacity to profoundly influence the opportunities that deep brain stimulation offers. A representative head frame system for stereotactic neurosurgery for the implantation of deep brain elec-trodes is shown in Fig. 5.1.

Another major consideration pertains to the energy source for the implantable pulse generator. Since the implantable pulse generator is implanted, the simplicity of replacing a battery is greatly complexified by the requirement for an associated surgical procedure. With the direct proportion of movement disorder to age, the risk of fully recovering from surgery is a matter of concern. Furthermore, the risk of

Fig. 5.1 A representative
stereotactic head frame
system for implanting deep
brain stimulation
electrodes provided by
Allegheny General
Hospital

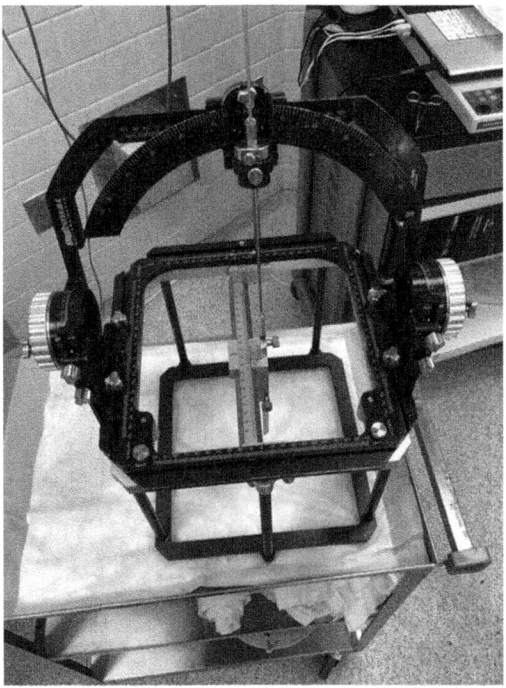

infection as a consequence of surgery for battery replacement impacts the design
considerations for a deep brain stimulation system.

Advances regarding the energy storage subsystem can significantly influence the
design of a deep brain stimulation system. A rechargeable battery scenario may
reduce the risk of additional surgical procedures. In addition, energy conservation
techniques, such as for optimizing the characteristics of the stimulation, may considerably protract the lifetime of the deep brain stimulation system in terms of available energy.

5.3   Deep Brain Stimulation Operative Risks and Complications

Intuitively since the procedure for implementation of a deep brain stimulation system requires neurosurgery, there exists a degree of inherent risk to the surgical
endeavor. With regard to the neurosurgery to implant the device, two potential issues
pertain to infection and intracranial hemorrhaging [1]. These scenarios are logically
correlated with the quality of the surgical environment.

During surgery, the possibilities to develop a hematoma or paralysis are complications of considerable harm to the patient. However, these high-risk complications
are considered rare in nature. More frequent complications that are likewise

detrimental are more attributed to the intrinsic quality of the deep brain stimulation hardware, such as the potential for infection and also fracture and dislocation of the deep brain stimulation electrode leads. The observation of such threats to a successful surgical outcome emphasizes the requirement for prudent screening of candidates for deep brain stimulation [3].

Four adverse effects that are attributed to the deep brain stimulation system hardware are:

- Infection
- Lead fracture
- Skin erosion
- Implantable pulse generator malfunction or lead migration

Although infrequent in nature, these issues impact patient morbidity. Therefore, their comprehension is instrumental for the determination of risk versus benefit of deep brain stimulation intervention [2].

Research has ascertained a disparity regarding infection rate for the preliminary surgery applying the electrodes. The straight-parasagittal incision infection rate exceeds the associated curvilinear incision procedure by roughly sixfold and was considered statistically significant. Treatment of infection can be resolved by antibiotics or in the more extreme situation the removal of the apparatus [2].

Fracture of the electrode leads can be determined through radiography. Another technique for diagnosing electrode lead fracture is through electrical examination. Treatment of this scenario requires replacement [2].

Skin erosion involves loss about the epidermis, which is the outer layer of skin. It can occur in the vicinity of the lead-extension connector, such as proximal to the mastoid. The symptoms can be alleviated through removal of the lead and extension with subsequent treatment of potential infection [2].

Implantable pulse generators can experience failed operation due to technical malfunction. The lead electrode migration is attributed to the electrode being insufficiently secured about the associated burr hole [2]. Techniques to resolve this issue are available, such as securing the lead electrode with bone cement [2, 9].

5.4   Postoperative Risk Regarding Energy Interaction

A matter of great concern regards the safety of the electrodes, especially with respect to an external energy source coupling with the electrodes to impart an adverse deposition of thermal energy. Electrosurgery systems and magnetic resonance imaging (MRI) scanner can generate high-frequency electric fields (RF) [4]. The inherent risk of the deep brain stimulation system is not only intrinsic to the novelty of this technology, but also due to the observation that the deep brain stimulation system interacts with neuroanatomy structures deep within the brain. Therefore, the potential for devastating contraindications is a sobering possibility in combination with other medical technologies. For example, consider the tandem application of diathermy.

A subject responding well to deep brain stimulation for the treatment of Parkinson's disease later underwent diathermy treatments following maxillary teeth extraction. The diathermy induction coils were set to conventional pulse and frequency settings. However, soon after the diathermy treatments, catastrophic neurological damage became clearly apparent [5].

The subject displayed considerable neurological deterioration. The implications of this rapid decline suggest tissue damage proximal to the subthalamic nucleus. Edema (fluid buildup) situated near the deep brain stimulation electrodes around the brainstem implied that the electrodes were heated due to radiofrequency current induction, since the diathermy device applied a pulse-modulated radiofrequency. A proposed mechanism was that the diathermy system couples with the electrode lead through radiofrequency current induction, which caused the electrodes to heat to the threshold of causing catastrophic neurological damage [5].

This unfortunate scenario emphasizes the considerable risk of two therapy interventions contraindicating with drastic and irreversible consequences. The observations underscore that extrinsic electric or magnetic fields of relative high energy can interact with electrode leads and other deep brain stimulation system components to induce considerable tissue damage proximal to the electrodes, which are located near highly delicate neuroanatomy structures deep within the brain [5]. In particular, the safety of MRI with its extreme magnetic intensity for subjects with deep brain stimulation is a subject that warrants further scrutiny.

Given the considerable amount of electromagnetic energy produced by an MRI system, there are multiple high-risk scenarios for applying MRI for a subject with an implanted deep brain stimulation system [6]. Potential hazards for the combination of MRI and deep brain stimulation are:

- Interactions caused by the magnetic field
- Disruption of the deep brain stimulation device's operation
- Electrical current produced from the electromagnetic energy
- Adverse heating, especially proximal to neurological structure [6]

The consequences are greatly amplified given the proximity of the deep brain stimulation system to highly sensitive structures of the brain.

Safety procedures are advised with respect to conducting MRI diagnostics for people with implanted deep brain stimulation devices. Regarding a person with a deep brain stimulation device, the specific absorption rate induced by the MRI system is recommended at a threshold of approximately an order of magnitude less than for people without deep brain stimulation. Additional safety procedures involve setting the deep brain stimulation device to "off" status, applying the amplitude parameter to 0.0 V, and also configuring the implantable pulse generator to a bipolar status. Another safety metric is the monitoring of current drain and impedance [7].

Postoperative determination of the implantation of the deep brain stimulation electrodes provides beneficial awareness. For example, the consideration of surgical complications, long-term circumstances, and more optimal tuning are feasible with regard to ascertaining the electrode implantation. An issue with MRI is the generation of substantial electromagnetic fields. These electromagnetic fields can produce dangerous interaction with the implanted aspects of the deep brain stimulation system. The benefits of this robust nonintrusive imaging technique for further establishing the efficacy of deep brain stimulation transcend the risks of applying MRI to the brain of a subject with implanted deep brain stimulation. Further research is advised to better determine the suitability of restrictions, such as the upper threshold for specific absorption rate [7].

5.5 Adverse Neurological and Neuropsychological Effects for Deep Brain Stimulation

There are also adverse outcomes for the application of deep brain stimulation. By contrast, the neurological and neuropsychological effects are broader in scope. Depending on their severity, these neurological and neuropsychological side effects may even warrant the discontinuation of deep brain stimulation therapy [1].

Neurological adverse effects pertaining to deep brain stimulation therapy are:

- Motor and sensory disturbance
- Disequilibrium
- Dysphagia
- Speech difficulty
- Cognitive impairment
- Memory deficits [1, 8]

Regarding the domain of neuropsychology, the side effects are:

- Suicidal ideation
- Mania
- Depression
- Anxiety [1, 8]

Awareness of these potential neurological and neuropsychological effects can better ascertain the long-term feasibility of utilizing deep brain stimulation for the therapeutic treatment of movement disorder symptoms.

5.6 Operative Technique Advocated by Allegheny General Hospital

The description of various operative techniques for deep brain stimulation is beyond the scope of this text. However, the surgical technique of Allegheny General Hospital is briefly summarized. Deep brain stimulation has traditionally been performed using a stereotactic head frame; however, frameless deep brain stimulation techniques have also become more popular. The protocol at Allegheny General Hospital has been to perform deep brain stimulation implantation in two stages: the first stage involves placement of the brain electrodes using a stereotactic head frame with the patient awake and the second stage involves placement of the extension wiring and pulse generators with the patient asleep under general anesthesia. The patient undergoing deep brain stimulation obtains a preoperative stereotactic MRI, for which target planning is performed. Indirect targeting is performed meaning that targets for deep brain stimulation, such as the subthalamic nucleus (STN), are identified by their location relative to the anterior commissure and posterior commissure. Microelectrode recording (MER) has been routinely performed to provide physiologic confirmation of lead location; however, the use of MER has been abandoned by some centers, and it has not been demonstrated that MER improves outcomes in deep brain stimulation. MER involves using a microelectrode and microdrive to listen to the neuronal firing patterns in the region of the target. MER recordings function as neuroanatomic fingerprints, which may assist with identifying the target and suggest how a change in trajectory might improve lead placement within the target [10].

During stage I, the patient undergoes placement of a Cosman-Roberts-Wells (CRW) stereotactic head frame (Integra Life Sciences, Plainsboro, NJ). The CRW frame uses an arc system which swivels a probe exocentrically around the target, so a fixed entry point is not necessary [10]. Stereotactic head frames for humans were first described in 1947 by Spiegel et al. [11], and they allow precise targeting of deep brain structures by rigidly merging the location of points in the brain to a Cartesian coordinate system [10].

The patient is kept awake during surgery, since sedation and general anesthesia would impair the ability to identify brain structures during microelectrode recording. For most stage I cases, bilateral deep brain stimulation electrodes are placed into the brain. For example, the deep brain stimulation electrode, which has four electrode contacts, the middle two of which are subdivided into three contacts at 120 degree angles to each other, permits directional deep brain stimulation programming [10].

During a standard deep brain stimulation surgery for Parkinson's disease, the bilateral STN is usually targeted, since this target has been associated with the ability to reduce anti-Parkinson's disease medications. The stage I surgery with bilateral STN lead placement typically takes about two hours. Anesthesia is requested to provide no sedatives, hypnotics, analgesics, or anesthetics before or during the procedure, since these drugs may reduce the quality and amplitude of the MER. Strict blood pressure control is important to reduce the risk of intracranial hemorrhage, and no lead, microelectrode, or cannulas are placed into the brain unless the systolic blood pressure is below 140 Hg [10].

During the surgery, bilateral MER is usually performed in order to confirm that at least 5 mm of STN has been traversed. In addition, macrostimulation is performed on each side to confirm that there is a wide therapeutic window. For macrostimulation, the permanent lead is placed and is connected to temporary wires controlled by a representative. Wide bipolar configuration is typically tested with the deepest contact being the cathode or negative contact and the most superficial contact being the anode or positive lead. Regarding this wide bipolar configuration, the voltage or current of the lead is increasingly raised while keeping the frequency and pulse width relatively fixed at 130 Hz and 60 microseconds, respectively [10].

For cases of deep brain stimulation for Parkinson's disease, the patient is examined during macrostimulation to detect changes in tremor and rigidity in the contralateral upper extremity. Moreover, care is taken to note any persistent side effects such as paresthesia, motor contractions, diplopia, mood changes, and speech changes. The findings through MER and macrostimulation help confirm the local environment for the electrode and may prompt the surgeon to reposition the electrode by as little as 1–2 mm in order to widen the therapeutic window [10].

Fibrin sealant glue is applied around the electrode in order to prevent cerebrospinal fluid (CSF) egress during the operation. CSF loss can lead to brain shift and resultant loss of targeting accuracy. After the electrode position has been optimized, the leads are locked in place with the burr hole cover locks, and the distal end of the lead is tunneled under the scalp in preparation for stage II of deep brain stimulation implantation [10].

Stage II of deep brain stimulation implantation is usually performed about one week later and involves a one and a half hour surgery where the extension wires are placed that connect the brain leads to the pulse generators. Stage II is performed under general anesthesia, and depending on the deep brain stimulation system used, either one or two pulse generators are implanted, usually in the subclavicular region below the collarbone [10].

In cases of deep brain stimulation for Parkinson's disease, patients have been routinely sent to inpatient rehabilitation after stage II. During inpatient rehabilitation, the process of medication reduction and initial deep brain stimulation programming can begin. Based on extensive experience, it can take six months or more to determine the optimal prescription of medications and deep brain stimulation parameter configuration settings for a particular patient with Parkinson's disease. Patients are given a controller that allows them to turn the deep brain stimulation "On" and "Off" and check pulse generator life. Patients may also be able to adjust some parameters of stimulation, if given this control by their surgeon [10].

5.7 Applied Deep Brain Stimulation Programming from by Allegheny General Hospital

Proper programming is crucial to ensure good clinical outcomes with deep brain stimulation. Deep brain stimulation programming remains an art, and there are no validated protocols or standardized algorithms which assist in the laborious process. Many centers wait two to four weeks after lead implantation to begin programming in order to allow the microlesion effect to wear off. The microlesion effect may engender temporary improvement of symptoms in the absence of stimulation and may be due to edema or microtrauma from the lead placement itself [10].

Deep brain stimulation programming often begins with a monopolar review in which each contact on a lead is tested as a cathode (negative contact) in monopolar mode with the case or pulse generator acting as the anode (positive contact). The main stimulation variables which are adjusted during programming include the pulse width, amplitude, and frequency of stimulation. Amplitude can either be expressed in volts or milliamperes depending on whether the pulse generator is voltage-controlled or current-controlled. From the monopolar review, the most beneficial contact is identified as well as thresholds for clinical benefit and thresholds for side effects. Double monopolar stimulation (two adjacent contact cathodes with pulse generator as anode) is considered in cases when single monopolar stimulation does not control motor symptoms, and bipolar modes (adjacent anode and cathode contacts on the lead) are often chosen if side effects are observed at lower amplitudes [12].

Deep brain stimulation programming requires specialized training, and patients often travel long distances to a center with clinicians trained to program deep brain stimulation. The art of programming involves finding the parameters which simultaneously optimize clinical benefit, minimize or eliminate side effects, and utilize the least amount of electrical energy. Traditional deep brain stimulation has been performed with voltage-controlled pulse generators that created continuous, spherical electrical fields around the lead contacts. Modern deep brain stimulation is incorporating more advanced programming features including interleaved programming (rapid, alternate activation of two contacts with different voltage and pulse widths but identical frequencies), current-controlled deep brain stimulation, intermittent deep brain stimulation, current fractionalization, and directional lead deep brain stimulation [10].

The objective for directional deep brain stimulation programming is that clinical benefit may be obtained with fewer side effects due to a larger therapeutic window. Furthermore, the lower energy requirement with directional programming is anticipated to extend pulse generator lifespan. Pulse generator replacement surgery is part of long-term maintenance therapy, and patients with high energy settings may require battery replacements as frequently as every one to two years. Newer-generation rechargeable pulse generators have increased battery lifespan to 15 years or more. However, this evolution of rechargeable pulse generators remains to be determined whether battery recharging is a significant burden for patients and whether this capability will affect long-term satisfaction with deep brain stimulation [10].

Given the complexity, expertise, travel, and time involving with deep brain stimulation programming, there has been increasing focus on developing closed-loop systems which program themselves by adjusting to signals generated by the patient [10]. One strategy is using feedback from brain signals, such as local field potentials (LFPs), recorded from the basal ganglia, such as the STN nucleus beta-band frequency [13]. Another closed-loop deep brain stimulation strategy would adjust deep brain stimulation in response to signals from external motion sensors that could capture real-time information on motor symptoms, such as tremor [14]. Finally, as deep brain stimulation matures as a therapy, it may become acceptable and feasible to permit patients to adjust their own programming settings via the use of home computers or even portable smart devices. This observation advocates the development of wearable and wireless systems for the treatment of neurodegenerative movement disorders, such as Parkinson's disease and Essential tremor, through the application of deep brain stimulation in the context of Network Centric Therapy.

In summary, deep brain stimulation has become a well-established, evidence-based therapy for movement disorders. In addition, deep brain stimulation has also established itself as a powerful tool for understanding brain neurophysiology in various disease states. The remarkable clinical results of deep brain stimulation have helped usher in a new era of neuroscience that describes neurological diseases in terms of disordered brain circuits instead of derangements of neurotransmitters, proteins, or genes [10].

5.8 Conclusion

Deep brain stimulation for the treatment of neurodegenerative movement disorders, such as Parkinson's disease and Essential tremor, provides considerable benefit for the appropriate patient. However, the surgery procedure is multiphased in application, and there are inherent risks, such as infection and potential for hemorrhaging. Nonobvious risks to the patient's health are relevant. For example, the energy derived from extrinsic electromagnetic fields can cause substantial harm, since the deep brain stimulation system is inherently an implanted device that extends to the structures of the deep brain. In light of this hazard, the benefits of utilizing MRI to evaluate quality and characteristics of the electrode insertion predominate when proper safety procedures are instilled. Furthermore, there are possible neurological and neuropsychological effects that can impact the quality of a patient's life that a clinician should be cognizant of throughout the long-term application of the deep brain stimulation system. Furthermore, as presented with applied demonstration, expert surgical techniques must be complimented with expert acuity during the parameter configuration tuning process. The rise of wearable and wireless systems is anticipated to facilitate the optimal application of deep brain stimulation for neurodegenerative movement disorders, such as Parkinson's disease and Essential tremor.

References

1. Okun MS (2012) Deep-brain stimulation for Parkinson's disease. N Engl J Med 367(16):1529–1538
2. Constantoyannis C, Berk C, Honey CR, Mendez I, Brownstone RM (2005) Reducing hardware-related complications of deep brain stimulation. Can J Neurol Sci 32(2):194–200
3. Hariz MI (2002) Complications of deep brain stimulation surgery. Mov Disord 17(S3):S162–S166
4. Patterson T, Stecker MM, Netherton BL (2007) Mechanisms of electrode induced injury. Part 2: clinical experience. Am J Electroneurodiagnostic Technol 47(2):93–113

5. Nutt JG, Anderson VC, Peacock JH, Hammerstad JP, Burchiel KJ (2001) DBS and diathermy interaction induces severe CNS damage. Neurology 56(10):1384–1386
6. Rezai AR, Phillips M, Baker KB, Sharan AD, Nyenhuis J, Tkach J, Henderson J, Shellock FG (2004) Neurostimulation system used for deep brain stimulation (DBS): MR safety issues and implications of failing to follow safety recommendations. Investig Radiol 39(5):300–303
7. Tagliati M, Jankovic J, Pagan F, Susatia F, Isaias IU, Okun MS (2009) Safety of MRI in patients with implanted deep brain stimulation devices. NeuroImage 47(S2):T53–T57
8. Temel Y (2010) Limbic effects of high-frequency stimulation of the subthalamic nucleus. Vitam Horm 82:47–63
9. Favre J, Taha JM, Steel T, Burchiel KJ (1996) Anchoring of deep brain stimulation electrodes using a microplate. Technical note. J Neurosurg 85(6):1181–1183
10. Tomycz ND, Whiting DM (2018) Deep brain stimulation: indications, operative technique, and programming. Internal Publication Allegheny General Hospital
11. Spiegel EA, Wycis HT, Marks M, Lee AJ (1947) Stereotaxic apparatus for operations on the human brain. Science 106(2754):349–350
12. Shukla AW, Zeilman P, Fernandez H, Bajwa JA, Mehanna R (2017) DBS programming: an evolving approach for patients with Parkinson's disease. Parkinson's Dis 8492619
13. Neumann WJ, Staub-Bertelt F, Horn A, Schanda J, Schneider GH, Brown P, Kuhn AA (2017) Long term correlation of subthalamic beta band activity with motor impairment in patients with Parkinson's disease. Clin Neurophysiol 128(11):2286–2291
14. LeMoyne R, Tomycz N, Mastroianni T, McCandless C, Cozza M, Peduto D (2015) Implementation of a smartphone wireless accelerometer platform for establishing deep brain stimulation treatment efficacy of essential tremor with machine learning. In: 37th Annual international conference of the IEEE, Engineering in Medicine and Biology Society (EMBS), pp 6772–6775

Chapter 6
Preliminary Wearable and Locally Wireless Systems for Quantification of Parkinson's Disease and Essential Tremor Characteristics

Abstract Inertial sensor systems, such as accelerometers, were proposed for the monitoring of human movement before their technology capability was sufficient for application to the human body. With sufficient progressive evolution, these sensors have been demonstrated for quantifying neurodegenerative movement disorders, such as Parkinson's disease and Essential tremor. Initial success was demonstrated for matters, such as medication efficacy and symptom status. Their further recent evolution has elucidated utility regarding the preliminary context of wearable and locally wireless systems. A novel configuration was proposed for the use of wearable and wireless accelerometer systems to provide quantified feedback to establish a strategy for acquiring optimal parameter settings for a deep brain stimulation system. Further demonstration of wearable and locally wireless inertial sensor systems for objectively quantifying neurodegenerative movement disorder tremor symptoms has been provided with local wireless connectivity to a proximally situated personal computer for post-processing. These developments establish the foundation for the extension to wearable and wireless inertial sensor systems with considerable accessibility to the Internet, such as provided by the smartphone. This foundation sets the precedence for the emergence of Network Centric Therapy regarding the domain of quantifying neurodegenerative movement disorders, such as Parkinson's disease and Essential tremor.

Keywords Movement disorder · Hand tremor · Parkinson's disease · Essential tremor · Quantification · Accelerometer · Gyroscope · Deep brain stimulation system · Optimal parameter configuration · Wearable and wireless system

6.1 Introduction

During the 1950s, inertial sensors, such as accelerometer systems, were proposed for the quantification of movement for human subjects [1–3]. However, during this stage, accelerometers were not sufficiently evolved, and they were even considered too cumbersome while also lacking sufficient reliability to monitor human

© Springer Nature Singapore Pte Ltd. 2019
R. LeMoyne et al., *Wearable and Wireless Systems for Healthcare II*,
Smart Sensors, Measurement and Instrumentation 31,
https://doi.org/10.1007/978-981-13-5808-1_6

movement. Influence from industries extrinsic to the biomedical community, such as automotive products and their airbag systems requiring feedback, evolved the miniaturization and acuity of the accelerometer system [2–4]. With the advent of wireless sensors, tethered devices have become obsolete [5]. Inertial sensors systems, such as accelerometers, have considerably evolved over the course of the past decade with transition from local wearable and wireless systems to demonstrate the preliminary development of the Internet of Things for the movement disorder community [2, 6–10].

An inherent advantage of applying wearable and wireless systems, such as accelerometers, for the quantification of movement disorder tremor is the opportunity to digitally store the signal data into a cumulative historical database. Trends pertaining to the timescale and rapidity of progression for a neurodegenerative movement disorder can be derived in a highly objective and quantified manner. Post-processing technique can refine the acuity of the clinician for optimal therapy strategy and possible transitions regarding therapy approaches [2, 6–10].

6.2   Preliminary Applications for Accelerometers Quantifying Parkinson's Disease

Accelerometer systems have been incorporated for the determination of medication intervention for the treatment of movement disorder tremor symptoms. Sophisticated numerical techniques, such as spectral analysis, have been applied characterize the respective acceleration waveforms. This integral technique has successfully determined the effectiveness of ameliorating tremor through medication [11].

Keijsers et al. investigated the efficacy of consolidating the accelerometer signals into a feature set for machine learning classification, such as a neural network, for distinguishing between Levodopa-induced dyskinesias (LID) contrasted to voluntary movement. The uniaxial accelerometers were aligned into orthogonal pairs and mounted about the wrist, upper arm, trunk, and leg. The accelerometer signal data was stored on a data recorder for post-processing. The neural network applied a multilayer perceptron, which consisted of an input layer, hidden layer, and output layer. The neural networks demonstrated the capacity to successfully differentiate LID compared to voluntary movements, and the machine learning technique also was capable of distinguishing severity of LID. Keijsers et al. propose their technique could enable the opportunity for automated evaluation in a quantified context [12].

Roughly a half decade later, Keijsers et al. developed an algorithm that differentiated between Parkinson's disease subjects respective of "On" and "Off" states. Parkinson's disease symptoms vary from "On" status that demonstrate Levodopa-induced dyskinesias (LID) to "Off" status that involve reappearance of Parkinson's disease symptoms. Patients are generally tasked with developing a diary to track their "On" and "Off" states for long-term assessment of the disease progression [13]. Intuitively a quantified methodology that incorporates wearable technology would be highly advantageous for the robust monitoring of "On" and "Off" states for people with Parkinson's disease.

Keijsers et al. applied six triaxial accelerometers mounted about the upper legs, upper arms proximal to the shoulders, trunk near the top of the sternum, and wrist respective of the most effected side. The accelerometer data was stored in a belt mounted data recorder secured about the subject's waist. The algorithm demonstrates the ability to distinguish between "On" and "Off" states for people with Parkinson's disease [13]. The research findings of Keijsers et al. infer the utility of wearable systems for the eventual autonomous objective and quantified diagnostic evaluation for people with movement disorder.

Scenarios presenting comorbidity can present a highly challenging medical intervention strategy. For example, a person with Parkinson's disease may also be afflicted with dementia. However, medical treatment of Parkinson's disease symptoms may be contraindicated by therapy intervention for dementia. For example, Rivastigmine is a medical intervention that can ameliorate the effects of cognitive affects for people with dementia. However, this intervention can also exacerbate the severity of Parkinson's disease tremor characteristics. Gurevich et al. applied accelerometry during 2006 to objectively quantify the nature of tremor severity in light of Rivastigmine treatment to enhance cognitive capability [14].

Gurevich et al. selected triaxial accelerometers to measure tremor severity. The accelerometers measured tremor amplitude in an objective and quantified manner for Parkinson's disease subjects during their "On" state. The application of this accelerometer technique enables the ability to develop an objectively quantified data history to contrast a baseline scenario to a longitudinal treatment through the medical intervention of Rivastigmine. Descriptive and inferential statistics may then be applied during the post-processing phase. In summary, Rivastigmine was determined to improve cognitive quality while minimally exacerbating tremor severity. Based on the context of these findings, Gurevich et al. advocate such medical intervention for advancing quality of life [14].

Obswegger et al. during 2001 provided preliminary demonstration of the utility of integrating a wearable accelerometer system mounted about the hand of the subject and deep brain stimulation for the treatment of movement disorders, such as Parkinson's disease and Essential tremor. The quantified data obtained from the accelerometer reveals notable improvement regarding movement disorder tremor with respect to the deep brain stimulation set to "On" compared to "Off" status [15]. Kumru et al. also applied a dorsum of the hand mounted wearable accelerometer strategy for the quantified contrast of movement characteristics involving deep brain stimulation set to "On" and "Off" mode [16].

A drawback to the original approach for applying wearable accelerometer systems for the measurement of human movement, such as tremor, is the manual nature of the data transfer. This methodology is summarized in Fig. 6.1. With the evaluation and progressive miniaturization of wireless applications, traditional data transfer strategies have become outmoded [5].

In particular LeMoyne et al. progressively evolved wearable and wireless accelerometer systems for an assortment of biomedical applications. Prior to the quanti-

Fig. 6.1 Experimental
approach for a wearable
body-mounted
accelerometer sensor
applying non-wireless
techniques, such as a data
logger for temporary
storage

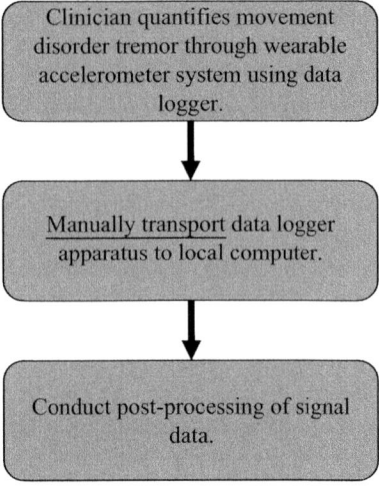

fication of movement disorder tremor, LeMoyne et al. established the validity of
wearable and wireless accelerometers for the quantification of reflex and gait. These
endeavors primarily featured a lightweight and volumetrically efficient wireless
accelerometer that could readily be worn about the leg of the subject [2, 6, 17–23].
Therefore, this application represents a preliminary wearable and wireless system.
Figure 6.2 presents the convenience of this preliminary configuration, which utilizes
local wireless connectivity.

The lightweight and compact nature of the effectively wearable and wireless
accelerometer system that LeMoyne et al. used for proof of concept for the capabilities regarding reflex and gait quantification was notably observable. Therefore, the
research team readily noted that the wireless accelerometer would be suitable in the
context of a wearable system for measuring movement disorder tremor about the
hand. Preliminary applications for Parkinson's disease hand tremor simulation
applied a flexible strap to secure the wearable and wireless accelerometer system
about the dorsum of the hand [24–26]. Later research, development, testing, and
evaluation of this strategy for quantifying simulated Parkinson's disease hand
tremor involved more robust mounting techniques, such as securing the wearable
and wireless accelerometer system to the dorsum of the hand through a glove [27].

6.3   Wireless Accelerometer Feedback for Optimal Tuning
of Deep Brain Stimulation Parameter Settings:
A Conceptual Perspective

During 2007, LeMoyne observed the inherent synergy between deep brain stimulation for the treatment of Parkinson's disease tremor symptoms and the role of wireless accelerometers for providing quantified feedback respective of its efficacy.

Fig. 6.2 Schematic process of a preliminary wearable and wireless accelerometer platform with wireless connectivity to a local PC

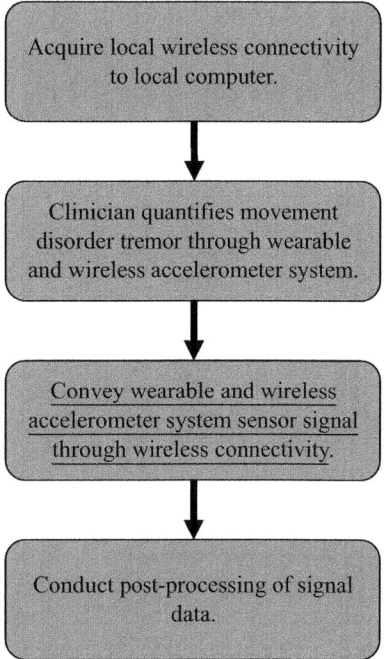

Acquire local wireless connectivity to local computer.

Clinician quantifies movement disorder tremor through wearable and wireless accelerometer system.

Convey wearable and wireless accelerometer system sensor signal through wireless connectivity.

Conduct post-processing of signal data.

Given myriad of parameter configurations, the optimal tuning of a deep brain stimulation system is nonobvious. Feedback through an ordinal methodology, such as the Unified Parkinson's Disease Rating Scale, is intuitively subjective. Three-dimensional wireless accelerometers can objectively quantify movement disorder characteristics from the temporal and frequency domain and also in terms of kinematic features, such as amplitude [24].

LeMoyne proposed the application of multidisciplinary design optimization for the assessment of local and globally optimal deep brain stimulation parameter settings with the objective of ascertaining maximal efficacy for mitigating the movement disorder symptoms associated with Parkinson's disease. This application would incorporate wireless accelerometers that are essentially wearable as a feedback mechanism [24]. Multidisciplinary design optimization has been successfully incorporated into complex engineering scenarios, such as modeling launch vehicle systems for economy and performance optimized access to space [28]. Variation in the design space topology may further elucidate the chronic progression of Parkinson's disease uniquely defined for the patient. The fundamental basis for gradient optimized neuromodulation was successfully presented as illustrated in Fig. 6.3 [24].

Fig. 6.3 Conceptual presentation of gradient optimized neuromodulation for acquiring optimal deep brain stimulation system parameter configuration with the dependent variable representing kinematic tremor features and the independent variables representing deep brain stimulation parameters [24]

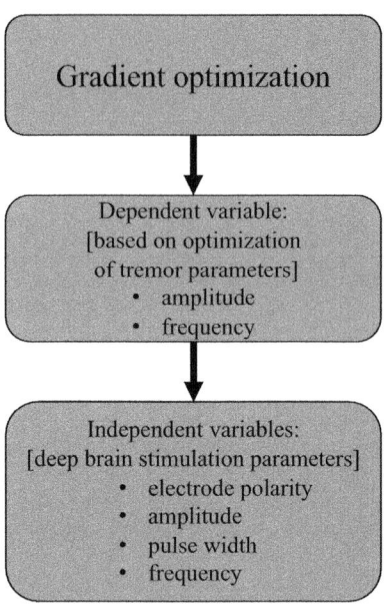

6.4   Preliminary Demonstration of Wearable and Wireless Accelerometer Systems for Quantifying Parkinson's Disease Tremor

The wireless three-dimensional accelerometer offers a robust alternative to the traditional ordinal scale strategies, such as the Unified Parkinson's Disease Rating Scale. Accelerometer signal data can offer objectively quantified feedback for the titration of drug therapy in addition to deep brain stimulation system parameter configurations. A notable advantage of this concept is the opportunity to measure a subject with Parkinson's disease at a time and location of their convenience, such as a homebound setting [25].

Using simulated Parkinson's disease hand tremor, the wearable and wireless accelerometer system demonstrated the capacity to objectively quantify hand tremor characteristics. The tremor simulation was contrasted to a statically positioned wireless accelerometer. The accelerometer signal was consolidated from three signals in three dimensions to a single acceleration magnitude signal through the application of the Pythagorean theorem. The data was further post-processed through acquiring the time-averaged acceleration of the static and simulated tremor signal for the acceleration magnitude. Using descriptive statistics, notable quantified contrast was observed [26].

A notable observation of this preliminary engineering proof of concept perspective for measuring simulated tremor is the capacity to further refine the methodology. From a purely experimental vantage, the energy constraints imposed due to the sampling rate of the accelerometer signal are not significant. However, for a robust homebound scenario, optimization of the energy demands for the accelerometer is warranted.

The research performed by LeMoyne et al. during 2009 applied wearable and wireless accelerometers that sampled at a rate of 2048 Hz [26]. The signal sampling rate may be substantially reduced in light of the Nyquist criterion which establishes a minimum necessary sampling threshold. Using the Nyquist criterion, the minimum sampling rate should be on the order of double the actual frequency [2, 6, 29–31].

The sampling rate of an accelerometer tasked to quantify tremor may be substantially reduced relative to the sampling rate applied by LeMoyne et al. during 2009, which was suited for experimental demonstration. Resting tremor for Parkinson's disease has been observed to occur at a rate of four to five per second [32, 33]. The implications are that a sampling threshold of 10 Hz would be appropriate. This enables a reduction of roughly two orders of magnitude.

Regarding the appropriateness of the minimum threshold, it is noted that a greater sampling rate may be appropriate in light of other desired extensions of the research. For example, the tandem sampling of six degrees of freedom through an accelerometer and gyroscope can enable sensor fusion, which can provide spatial representation (displacement, velocity, and acceleration) as a function of time for the movement disorder tremor being quantified. In order to provide robust sensor fusion, sampling at a rate on the order of 1000 Hz for enabling a proper orientation filter, such as a Kalman filter, is prudent [34].

The observation of competing requirements for establishing sampling rates of inertial sensors to quantify movement disorder tremor underscores the importance of clearly defining system requirements. If the object is to solely optimize sampling rate for quantifying tremor with regard to battery lifetime and inherent energy conservation, then sampling at 10 Hz is appropriate. However, if the system objective is to provide both quantification of tremor through inertial sensors and sensor fusion for spatial-temporal definition, a sampling of 1000 Hz may be necessary.

Also, during 2009, Giuffrida et al. presented a tremor quantification device known as Kinesia™. This application represents a wearable and wireless system, and motion sensor data is transmitted wirelessly from the command module. Inertial sensor data is acquired through accelerometer and gyroscope signals using a ring-mounted device. The data stream is conveyed by wire to the command module that is secured to the patient's wrist for later transmission for post-processing [35].

The device developed by Giuffrida et al. demonstrates a unique conceptual architecture. This system applies a relatively small inertial sensor positioned about the finger as a ring [35]. In essence, the research team of Giuffrida is seizing upon the opportunities made available regarding miniaturization for wearable sensor systems.

However, respective of the time of development Kinesia™ applied a small locally wired tethering to a larger command module. The command module utilizes wireless connectivity to convey its data for post-processing [35]. An issue for this state of development is the potential for the tethered wiring connecting the ring worn inertial sensor to the wireless command module to become caught during

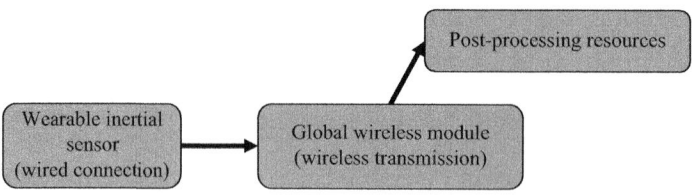

Fig. 6.4 Perspective architecture of a wearable inertial sensor locally conveying the data stream through wiring to a relatively global wireless module for transmission to subsequent post-processing resources

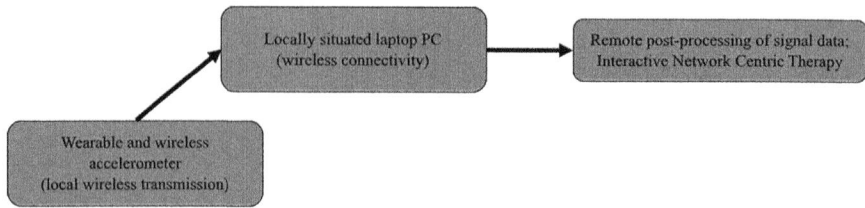

Fig. 6.5 Conceptual architecture for the data transmission using a wearable and wireless accelerometer node through local wireless connectivity to a laptop PC for post-processing

daily activities and causing discontinuity respective of the inertial sensor data stream. A perspective illustration of this novel architecture for wearable wired to wireless sensor system is presented in Fig. 6.4. As detailed later in Chap. 7, "Wearable and wireless systems with Internet connectivity for quantification of Parkinson's disease and Essential tremor characteristics," advances in inertial sensor nodes with local Bluetooth wireless connectivity to a more globally wireless smartphone, portable media device, or tablet enable a considerable advance by contrast to this preliminary architecture.

During 2013, LeMoyne et al. researched, developed, tested, and evaluated a wearable and wireless accelerometer system with local wireless connectivity to a laptop PC. The conceptual architecture for this methodology is presented in the schematic represented in Fig. 6.5. The wireless accelerometer node was mounted about the dorsum of the hand and secured through a glove. The experimental objective was to distinguish through quantified contrast a statically positioned wireless accelerometer node (control) to a glove-mounted wireless accelerometer node (experimental) representing simulated Parkinson's disease hand tremor. The equipment is illustrated in Fig. 6.6 [27].

Relative to the experiment conducted by LeMoyne et al. during 2009, the sampling rate was reduced to 512 Hz. The implication was a reduced time for transmission of the data package from the wireless accelerometer node to the local laptop PC. Wireless connectivity to the laptop PC for post-processing the accelerometer signal data was enabled by a wireless base station with a USB connector [27].

The accelerometer signal data was consolidated from the orthogonal three-axis accelerometer vectors to an acceleration magnitude. Time averaging of the acceleration magnitude for the static condition and experimental simulated tremor was cal-

Fig. 6.6   Wearable and wireless accelerometer system (experimental) with glove mounted experimental simulated Parkinson's disease hand tremor and wireless accelerometer system (control) that is statically mounted [27]

culated for further post-processing. Figure 6.7 illustrates the experimental acceleration magnitude signal for the simulated Parkinson's disease hand tremor, for which the control static acceleration magnitude signal revealed in Fig. 6.8 is notably disparate. Descriptive statistics derived the mean, standard deviation, and coefficient of variation. Inferential statistic using a one-way ANOVA with alpha <0.05 contrasted the simulated Parkinson's disease hand tremor to the static condition, and statistical significance was achieved. The application of the wearable and wireless accelerometer mounted by glove for the quantification of simulated Parkinson's disease hand tremor demonstrates considerable degree of accuracy, consistency, and reliability [27].

6.5   Evolution of Wearable and Wireless Systems: From Local Wireless Connectivity to Internet Connectivity

Wearable and locally wireless systems for the objective quantification of neurodegenerative movement disorders, such as Parkinson's disease and Essential tremor, establish the technology evolution pathway for the development of Network Centric Therapy. Although inertial sensor devices have been proposed for recording the features of movement disorder tremor symptoms, their progressive improvement needed to reach a threshold for sufficient miniaturization, in order to be properly mounted to an aspect of the human anatomy, such as the dorsum of the hand.

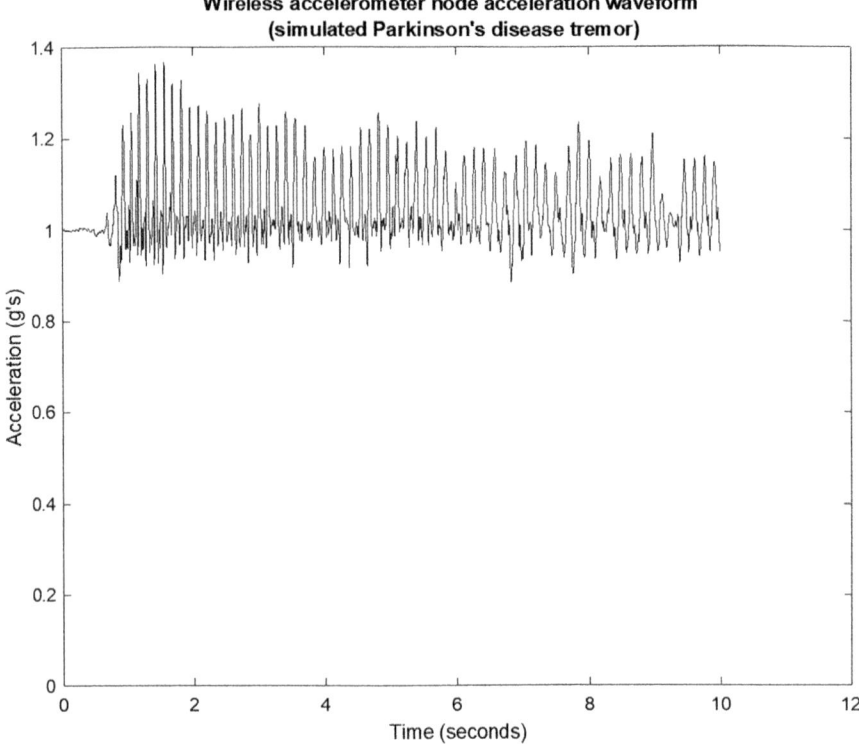

Fig. 6.7 Acceleration magnitude signal (experimental) of the simulated Parkinson's disease tremor [27]

Advances regarding wireless technology eventually enabled local connectivity to be nearly situated relative to computational resources for data signal post-processing.

In essence, the amalgamation of these capabilities constitutes the nascent origins of Network Centric Therapy. The preliminary foundation of Network Centric Therapy is presented in Fig. 6.9. In this configuration, the signal data of the inertial sensor is conveyed through locally wireless connectivity to a proximally situated personal computer. With the personal computer connected to the Internet, Network Centric Therapy is feasible for which the data package can be post-processed in a remotely situated context. This conceptual architecture is transcended by the rampant emergence of wearable and wireless systems, such as the smartphone. In Chap. 7, "Wearable and wireless systems with Internet connectivity for quantification of Parkinson's disease and Essential tremor characteristics," Internet connectivity is achieved through wireless connectivity to the Internet directly on behalf of the wearable and wireless system itself. This development further expands the utility and opportunity for the evolution of Network Centric Therapy.

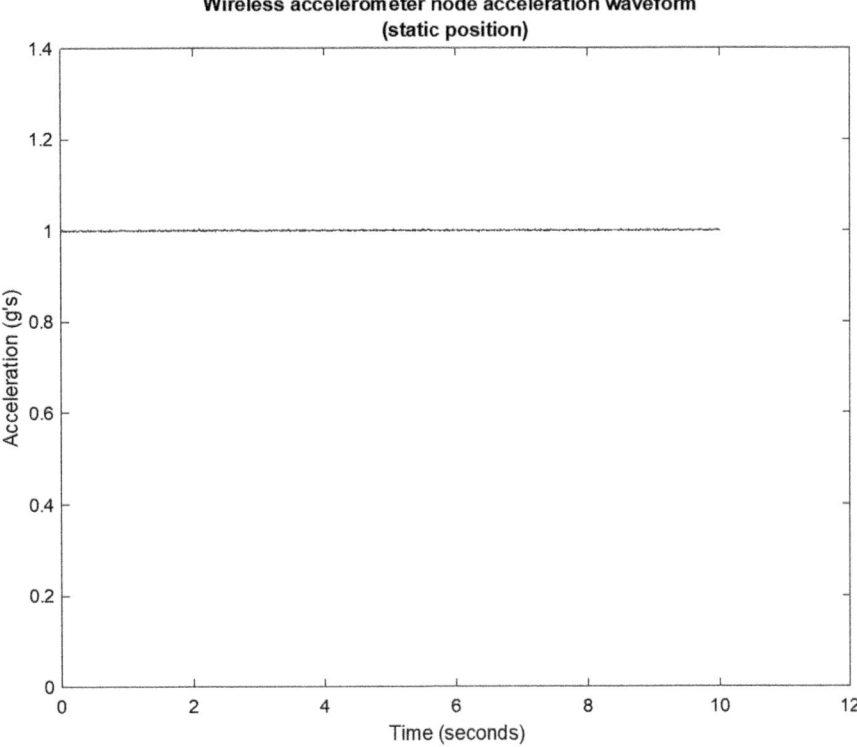

Fig. 6.8 Acceleration magnitude signal (control) of the statically situated wireless accelerometer [27]

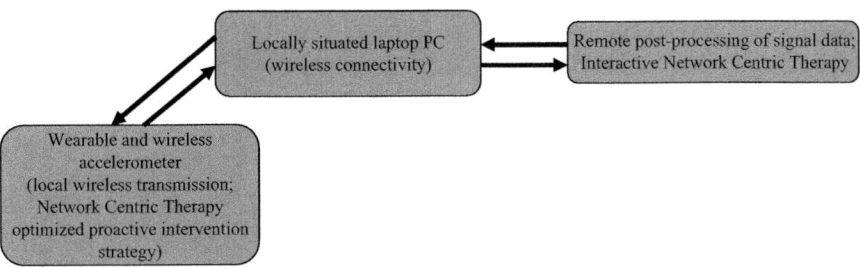

Fig. 6.9 Schematic process of a preliminary foundation for Network Centric Therapy configuration with wearable and locally wireless systems achieving Internet connectivity through a local computer

6.6   Conclusion

Inertial sensor systems have been proposed for the quantification of human movement prior to the technology being feasible. The progressive evolution of inertial sensors, such as miniaturization, leads to the application of wearable inertial sensors to quantify neurodegenerative movement disorders, such as Parkinson's disease and Essential tremor. These wearable systems enabled the quantification of medication efficacy and symptom status from an objective frame of reference. With evolutionary pathways for associated technologies, the capability to incorporate wearable and locally wireless inertial sensor systems for the quantification of neurodegenerative movement disorders, such as Parkinson's disease and Essential tremor, was realized.

Local wireless connectivity was achieved with the use of a proximally situated personal computer to receive the signal data. Furthermore, with the emergence of wearable and locally wireless systems as feedback for the efficacy of deep brain stimulation, a conceptual architecture using a robust optimization technique was proposed for the determination of the most efficacious deep brain stimulation parameter configuration. These preliminary trends regarding the development of wearable and wireless systems for neurodegenerative movement disorders, such as Parkinson's disease and Essential tremor, establish the foundation for Network Centric Therapy.

Future developments relative to the presence of wearable and locally wireless systems advocate the evolution to wearable and wireless systems that can achieve wireless connectivity to the Internet based on its own intrinsic capabilities. The smartphone constitutes a wearable and wireless inertial sensor platform thoroughly discussed in Chap. 7 "Wearable and wireless systems with Internet connectivity for quantification of Parkinson's disease and Essential tremor characteristics." These progressively evolutionary increments further realize the considerable potential of Network Centric Therapy for providing optimal treatment for neurodegenerative movement disorders, such as Parkinson's disease and Essential tremor.

References

1. Saunders JB, Inman VT, Eberhart HD (1953) The major determinants in normal and pathological gait. J Bone Joint Surg 35A(3):543–558
2. LeMoyne R, Coroian C, Cozza M, Opalinski P, Mastroianni T, Grundfest W (2009) The merits of artificial proprioception, with applications in biofeedback gait rehabilitation concepts and movement disorder characterization. In: Biomedical engineering. Intech, Vienna, pp 165–198
3. LeMoyne R, Coroian C, Mastroianni T, Grundfest W (2008) Accelerometers for quantification of gait and movement disorders: a perspective review. J Mech Med Biol 8(2):137–152
4. Culhane KM, O'Connor M, Lyons D, Lyons GM (2005) Accelerometers in rehabilitation medicine for older adults. Age Ageing 34(6):556–560
5. Patel S, Park H, Bonato P, Chan L, Rodgers M (2012) A review of wearable sensors and systems with application in rehabilitation. J Neuroeng Rehabil 9(1):21
6. LeMoyne R, Mastroianni T (2018) Wearable and wireless systems for healthcare I: gait and reflex response quantification. Springer, Singapore

7. LeMoyne R, Mastroianni T (2015) Use of smartphones and portable media devices for quantifying human movement characteristics of gait, tendon reflex response, and Parkinson's disease hand tremor. In: Mobile health technologies, methods and protocols. Springer, New York, pp 335–358
8. LeMoyne R, Mastroianni T (2017) Wearable and wireless gait analysis platforms: smartphones and portable media devices. In: Wireless MEMS networks and applications. Elsevier, New York, pp 129–152
9. LeMoyne R, Mastroianni T (2016) Telemedicine perspectives for wearable and wireless applications serving the domain of neurorehabilitation and movement disorder treatment. In: Telemedicine, SMGroup, Dover, Delaware, pp 1–10
10. LeMoyne R, Mastroianni T (2017) Smartphone and portable media device: a novel pathway toward the diagnostic characterization of human movement. In: Smartphones from an applied research perspective. InTech, Rijeka, Croatia, pp 1–24
11. Schrag A, Schelosky L, Scholz U, Poewe W (1999) Reduction of parkinsonian signs in patients with Parkinson's disease by dopaminergic versus anticholinergic single-dose challenges. Mov Disord 14(2):252–255
12. Keijsers NL, Horstink MW, van Hilten JJ, Hoff JI, Gielen CC (2000) Detection and assessment of the severity of levodopa-induced dyskinesia in patients with Parkinson's disease by neural networks. Mov Disord 15(6):1104–1111
13. Keijsers NL, Horstink MW, Gielen SC (2006) Ambulatory motor assessment in Parkinson's disease. Mov Disord 21(1):34–44
14. Gurevich TY, Shabtai H, Korczyn AD, Simon ES, Giladi N (2006) Effect of rivastigmine on tremor in patients with Parkinson's disease and dementia. Mov Disord 21(10):1663–1666
15. Obwegeser AA, Uitti RJ, Witte RJ, Lucas JA, Turk MF, Wharen RE Jr (2001) Quantitative and qualitative outcome measures after thalamic deep brain stimulation to treat disabling tremors. Neurosurgery 48(2):274–284
16. Kumru H, Summerfield C, Valldeoriola F, Valls-Solé J (2004) Effects of subthalamic nucleus stimulation on characteristics of EMG activity underlying reaction time in Parkinson's disease. Mov Disord 19(1):94–100
17. LeMoyne R, Dabiri F, Coroian C, Mastroianni T, Grundfest W (2007) Quantified deep tendon reflex device for assessing response and latency. In: 37th Society for Neuroscience annual meeting
18. LeMoyne R, Coroian C, Mastroianni T, Grundfest W (2008) Quantified deep tendon reflex device for response and latency, third generation. J Mech Med Biol 8(4):491–506
19. LeMoyne R, Mastroianni T, Kale H, Luna J, Stewart J, Elliot S, Bryan F, Coroian C, Grundfest W (2011) Fourth generation wireless reflex quantification system for acquiring tendon reflex response and latency. J Mech Med Biol 11(1):31–54
20. LeMoyne RC (2010) Wireless quantified reflex device. Ph.D. Dissertation UCLA
21. LeMoyne R, Mastroianni T, Coroian C, Grundfest W (2011) Tendon reflex and strategies for quantification, with novel methods incorporating wireless accelerometer reflex quantification devices, a perspective review. J Mech Med Biol 11(3):471–513
22. LeMoyne R, Coroian C, Mastroianni T, Grundfest W (2009) Wireless accelerometer assessment of gait for quantified disparity of hemiparetic locomotion. J Mech Med Biol 9(3):329–343
23. LeMoyne R, Coroian C, Mastroianni T, Grundfest W (2008) Virtual proprioception. J Mech Med Biol 8(3):317–338
24. LeMoyne R (2007) Gradient optimized neuromodulation for Parkinson's disease. In: 12th Annual UCLA research conference on aging
25. LeMoyne R, Coroian C, Mastroianni T (2008) 3D wireless accelerometer characterization of Parkinson's disease status. In: Plasticity and repair in neurodegenerative disorders (Conference)
26. LeMoyne R, Coroian C, Mastroianni T (2009) Quantification of Parkinson's disease characteristics using wireless accelerometers. In: ICME International conference on IEEE Complex Medical Engineering (CME), pp 1–5
27. LeMoyne R, Mastroianni T, Grundfest W (2013) Wireless accelerometer configuration for monitoring Parkinson's disease hand tremor. Adv Park Dis 2(2):62–67

28. LeMoyne R (2008) Multidisciplinary cost and performance optimization of a two stage liquid propulsion based launch vehicle. In: 15th AIAA international space planes and hypersonic systems and technologies conference, AIAA 2008–2642
29. LeMoyne R (2016) Advances for prosthetic technology: from historical perspective to current status to future application. Springer, Tokyo
30. LeMoyne R (2016) Testing and evaluation strategies for the powered prosthesis, a global perspective. In: Advances for prosthetic technology: from historical perspective to current status to future application. Springer, Tokyo, pp 37–58
31. LeMoyne R, Mastroianni T (2018) Quantification systems appropriate for a clinical setting. In: Wearable and wireless systems for healthcare I: gait and reflex response quantification. Springer, Singapore, pp 31–44
32. Kandel ER, Schwartz JH, Jessell TM (2000) Principles of neural science. McGraw-Hill, New York, Ch 43
33. Bickley LS, Szilagyi PG (2003) Bates' guide to physical examination and history taking. Lippincott Williams and Wilkins, Philadelphia, Ch 16
34. LeMoyne R, Mastroianni T (2018) Quantifying the spatial position representation of gait through sensor fusion. In: Wearable and wireless systems for healthcare I: gait and reflex response quantification. Springer, Singapore, pp 105–110
35. Giuffrida JP, Riley DE, Maddux BN, Heldman DA (2009) Clinically deployable Kinesia™ technology for automated tremor assessment. Mov Disord 24(5):723–730

Chapter 7
Wearable and Wireless Systems with Internet Connectivity for Quantification of Parkinson's Disease and Essential Tremor Characteristics

Abstract Wearable and wireless systems for the objective quantification of neuro-degenerative movement disorder status, such as Parkinson's disease, have been successful achieved through the application of a smartphone. Preliminarily, the smartphone represented a wearable and wireless accelerometer system, which could be readily mounted to the dorsum of the hand through a glove. The initial proof-of-concept demonstration had broad implications. The experimental and post-processing resources were situated on effectively opposite sides of the continental United States of America. Through the smartphone's wireless connectivity to the Internet, the post-processing resources to reduce the data and the experimentation sited could be located effectively anywhere in the world. Furthermore, the experimental location could be selected based on the patient's preference. Another exemplary wearable and wireless system is the portable media device. As an extension of this wearable and wireless system capability, the smartphone was successfully applied to ascertain from a quantified perspective the efficacy of deep brain stimulation for Essential tremor. Extrapolations of inertial signal data for a wearable and wireless system, such as a smartphone, advocate the application of machine learning classification to distinguish between deep brain stimulation efficacy regarding "On" and "Off" status. Future evolutions of wearable and wireless systems for the objective quantification of neurodegenerative movement disorder status, such as Parkinson's disease and Essential tremor, underscore the value of local wireless connectivity from an inertial sensor node to a more powerful wireless system, such as a smartphone or tablet, to achieve Internet connectivity. These trends provide preliminary realization of the opportunities that Network Centric Therapy can enable with inertial sensor signal data stored in a Cloud computing database for post-processing to achieve patient-specific intervention and optimized deep brain stimulation parameter configurations.

Keywords Wearable and wireless system · Smartphone · Portable media device · Smartwatch · Tablet · Wireless Internet connectivity · Bluetooth wireless · Inertial sensor · Accelerometer · Gyroscope · Parkinson's disease · Essential tremor

© Springer Nature Singapore Pte Ltd. 2019
R. LeMoyne et al., *Wearable and Wireless Systems for Healthcare II*,
Smart Sensors, Measurement and Instrumentation 31,
https://doi.org/10.1007/978-981-13-5808-1_7

7.1   Introduction

The intrinsic characteristics of the smartphone enable the application as an initial wearable and wireless system for the objective quantification of movement disorder symptoms, such as hand tremor for Parkinson's disease. The experimental findings revealed that inertial sensor signal data for quantifying tremor could be wirelessly conveyed to the Internet as an email attachment. Also, the smartphone was effectively temporarily wearable through mounting the smartphone to the dorsum of the hand. Furthermore, the experimental site could be selected as the preference of the patient with post-processing resources situated anywhere in the world. The findings can be readily extended to applications of quantifying an assortment of human movement characteristics, such as movement disorder tremor, through the application of smartphones, portable media devices, and other wireless inertial sensors [1–11].

Further research, development, testing, and evaluation of wearable and wireless systems successfully extended toward the domain of Parkinson's disease and Essential tremor for deep brain stimulation. The application of a smartphone to establish the efficacy of deep brain stimulation for treating Parkinson's disease and Essential tremor was ascertained from a quantified perspective using a wearable and wireless system, such as a smartphone, with respect to "On" and "Off" status. An extension to the field of machine learning for classifying deep brain stimulation response has been proposed and later successfully demonstrated [12–15].

There are other novel schemes for the application of wearable and wireless systems for the evaluation of movement disorder symptoms. For example, the actual wearable and wireless inertial sensor could be further minimized through utilizing local Bluetooth wireless connectivity with the data package conveyed to the Internet through more global wireless connectivity achieved by a smartphone or tablet [16]. In summary, these preliminary demonstrations of wearable and wireless systems advocate the prevalence of Network Centric Therapy for the optimal patient-specific treatment intervention of neurodegenerative movement disorders, such as Parkinson's disease and Essential tremor.

7.2   Smartphone for Quantifying Parkinson's Disease Hand Tremor

The smartphone is comprised of an inertial sensor for quantifying movement. Furthermore, its wireless connectivity for accessing the Internet is on the order of a telecommunications footprint. A contextually specific application can be developed for recording a prescribed sample duration of a predetermined inertial sensor, such as the accelerometer. The experimental trial data could then be conveyed by wireless connectivity to the Internet as an email attachment, which resembles the functionality of a Cloud computing database, for post-processing at a remote location.

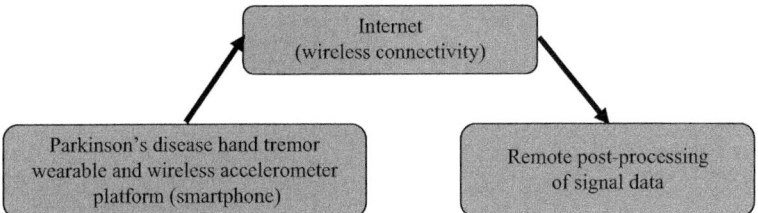

Fig. 7.1 Conceptual architecture for recording Parkinson's disease hand tremor through application of a wearable and wireless accelerometer platform, such as a smartphone, with experimental and post-processing resources remotely situated

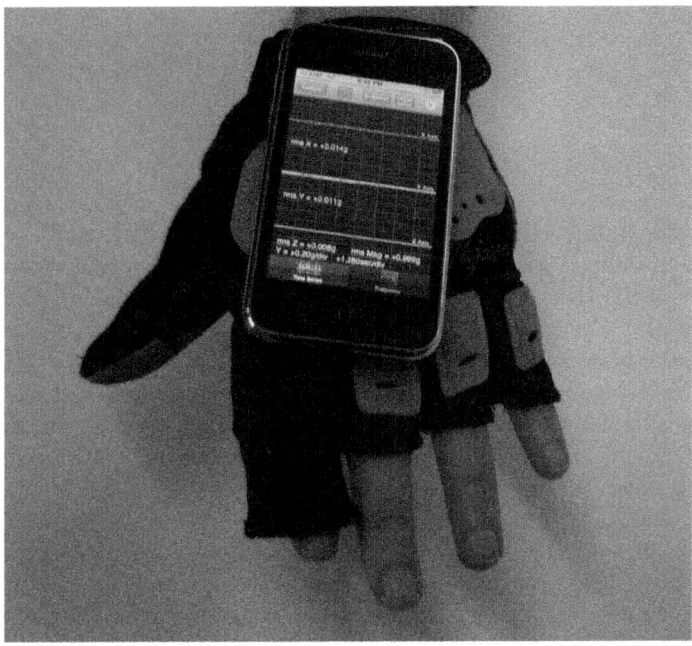

Fig. 7.2 Smartphone mounted about the dorsum of the hand as a wearable and wireless accelerometer platform for quantifying Parkinson's disease hand tremor [7]

In particular, the smartphone functioning as a wireless accelerometer platform is well suited to be mounted about the dorsum of the hand for quantifying Parkinson's disease hand tremor. The implications of this architecture for acquiring Parkinson's disease hand tremor are presented in Fig. 7.1. In summary, the experimental site using the smartphone as a wearable and wireless accelerometer platform can be conducted at a location of the patient's preference, and the post-processing resource can reduce the data anywhere in the world [1–6].

As an extension to the technology capabilities of wearable and wireless systems for quantifying Parkinson's disease tremor of the hand, the smartphone presents a logical extrapolation. As illustrated in Fig. 7.2, the smartphone can be readily

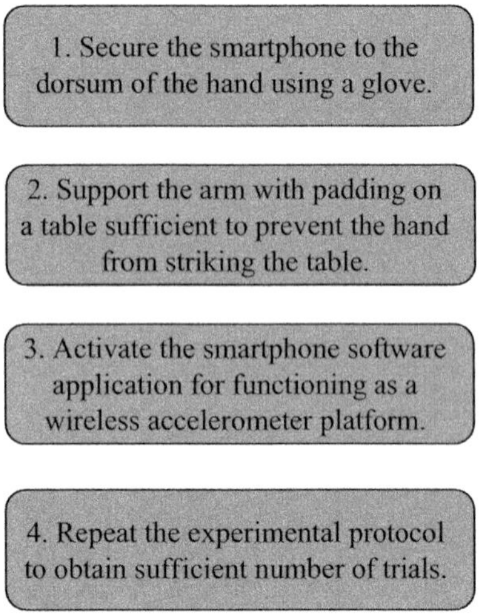

Fig. 7.3 Experimental protocol for the operation of a smartphone as a wearable and wireless accelerometer platform to quantify Parkinson's disease hand tremor [7]

1. Secure the smartphone to the dorsum of the hand using a glove.

2. Support the arm with padding on a table sufficient to prevent the hand from striking the table.

3. Activate the smartphone software application for functioning as a wireless accelerometer platform.

4. Repeat the experimental protocol to obtain sufficient number of trials.

mounted about the hand through a glove. A software application enables the smartphone to function as a wearable and wireless accelerometer platform with a sampling rate of 100 Hz and recording duration of 10 s. The trial data can be wirelessly transmitted as an email attachment for subsequent post-processing. A representative experimental protocol is demonstrated in Fig. 7.3 for the acquisition of Parkinson's disease hand tremor through a smartphone functioning as a wearable and wireless accelerometer platform. For preliminary proof of concept from an engineering perspective, a subject with Parkinson's disease is contrasted to a subject without Parkinson's disease [7].

The first stage for post-processing the accelerometer data was to consolidate the three orthogonal acceleration signals to the acceleration magnitude through application of the Pythagorean theorem. The original orthogonal acceleration signals derived from the smartphone accelerometer are presented in Fig. 7.4 for the subject with Parkinson's disease and Fig. 7.5 for the subject without Parkinson's disease. Figures 7.6 and 7.7 illustrate the consolidated acceleration waveform of the accelerometer magnitude for the subject with Parkinson's disease and the subject without Parkinson's disease. The time-averaged acceleration of the acceleration magnitude signals for the subject with Parkinson's disease hand tremor was contrasted to the subject without Parkinson's disease with regard to descriptive statistics in terms of mean, standard deviation, and coefficient of variation [7].

The notable quantified contrast of the descriptive statistics for the subject with Parkinson's disease and without Parkinson's disease warranted the application of inferential statistics for further analysis of the respective acceleration magnitude waveforms. The subject with Parkinson's disease and subject without Parkinson's disease were interpreted as two independent groups. Therefore, a two-group independent t-test with unequal variances was applied with alpha <0.05. Statistical significance was achieved with respect to the two subjects [7].

Fig. 7.4 Smartphone recording of the orthogonal acceleration signals for the subject with Parkinson's disease [7]

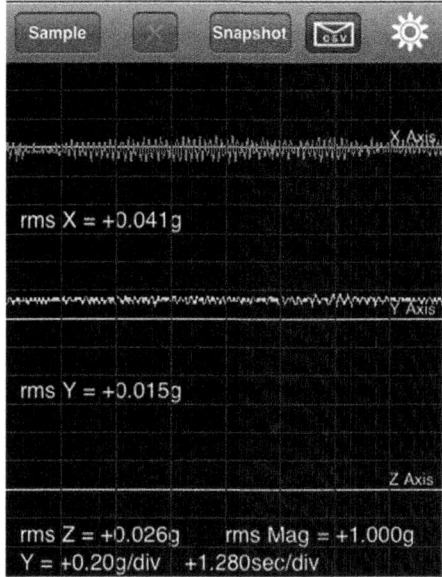

Fig. 7.5 Smartphone recording of the orthogonal acceleration signals for the subject without Parkinson's disease [7]

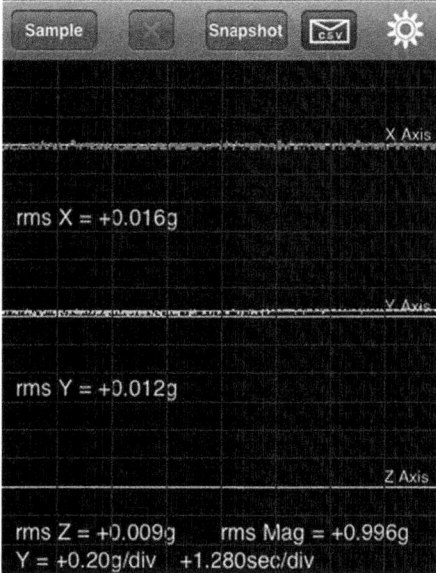

Further analysis of the acceleration magnitude signal can be conducted by considering the frequency domain. Figures 7.8 and 7.9 represent the frequency domain of the acceleration magnitude signal for the subject with Parkinson's disease and without Parkinson's disease. Note that the frequency signature for the Parkinson's

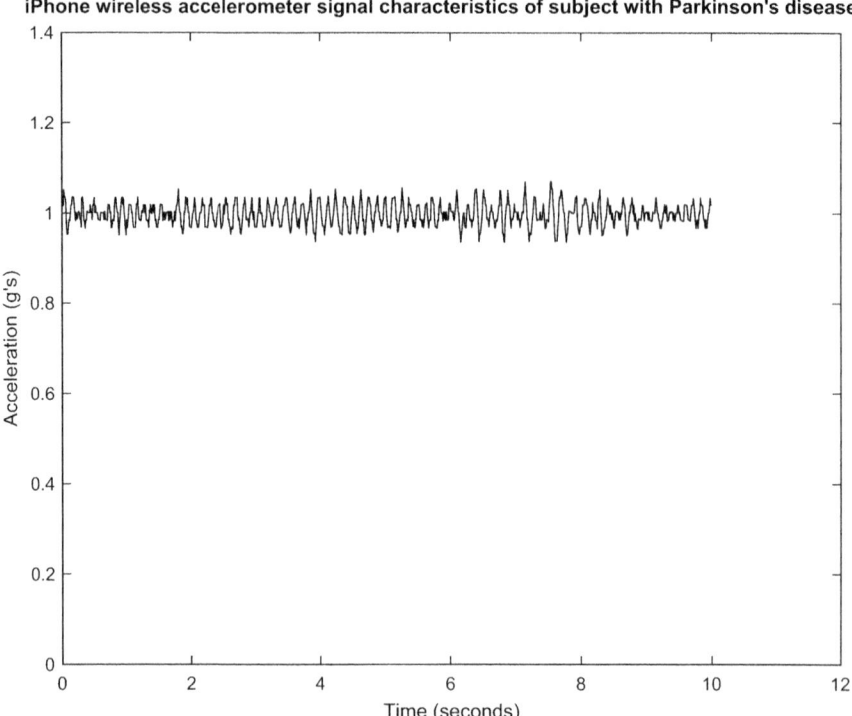

Fig. 7.6 Post-processed accelerometer signal data acquired by smartphone consolidated into an acceleration magnitude signal for the subject with Parkinson's disease [7]

disease subject is perceptively disparate when contrasted to the subject without Parkinson's disease [7].

There were considerably broad implications revealed upon this successful experimental endeavor. The capability to wirelessly transmit experimental data packages as conveniently attached email to the site of the research team's choosing through a smartphone functioning as a wearable and wireless accelerometer platform demonstrates the broad capabilities for the increasingly prevalent Internet of Things for the healthcare domain. With regard to the specific experiment itself, the experimental data was acquired in Pittsburgh, Pennsylvania, and conveyed through wireless connectivity to the Internet to Los Angeles, California, which effectively spans the continental United States. Therefore, given the omnipresent nature of the Internet, the experimental site and post-processing resources can literally be situated anywhere in the world [1–7].

On the order of a year after the findings of LeMoyne et al. during 2010, Kostikis et al. during 2011 developed similar smartphone system for quantifying Parkinson's disease tremor. Kostikis et al. applied the smartphone roughly orthogonal to the bones for the dorsum of hand, and the finger portions of the glove were cut for expanded dexterity. The device utilizes both the accelerometer and gyroscope

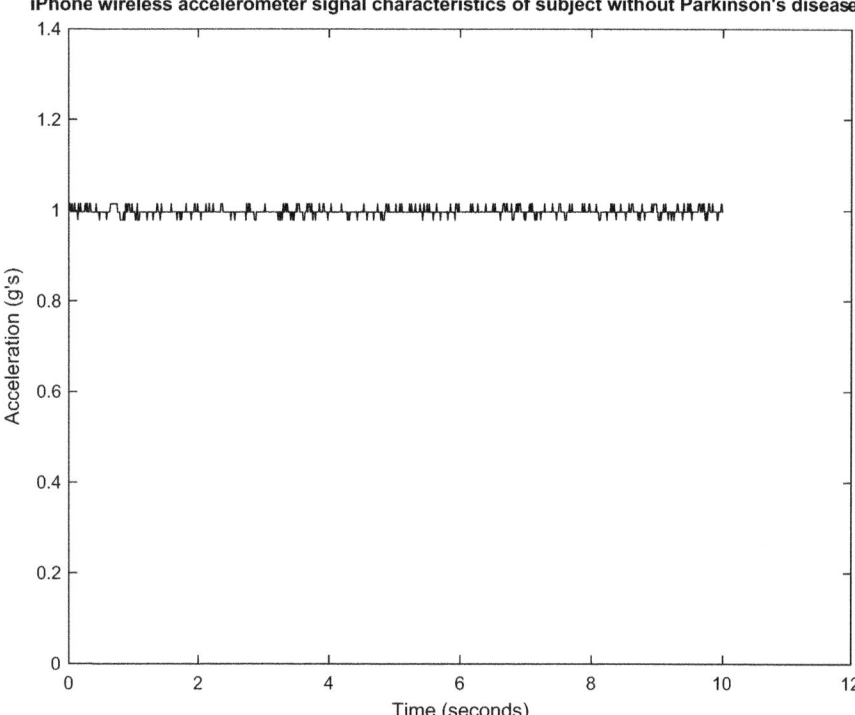

Fig. 7.7 Post-processed accelerometer signal data acquired by smartphone consolidated into an acceleration magnitude signal for the subject without Parkinson's disease [7]

signals to successfully distinguish between subjects with and without movement disorder. Kostikis et al. further develop their concept into a web-based application for enhance user interaction [8].

The portable media device is another wearable and wireless system capable for quantifying movement disorder, such as hand tremor, given its inertial sensor package. The smartphone, such as an iPhone, and portable media device, such as an iPod, utilize the same operating software, so they can apply the same application that permits them to function as wearable and wireless inertial sensor platforms. The appropriateness of selecting a smartphone or a portable media device is highly dependent on the environmental conditions, for which the experimentation is planned to be conducted [1, 3–6, 9–11].

The major disparities for selecting either a portable media device or a smartphone for quantifying movement disorder pertain to the wireless coverage range and cost. Although the portable media device is restricted to a local wireless zone, the portable media device only involves a singular fixed cost for purchase. For general areas with local wireless connectivity to the Internet, the portable media device may represent a preferable wearable and wireless system for the quantification of movement disorder, especially if multiple devices are required. By contrast the

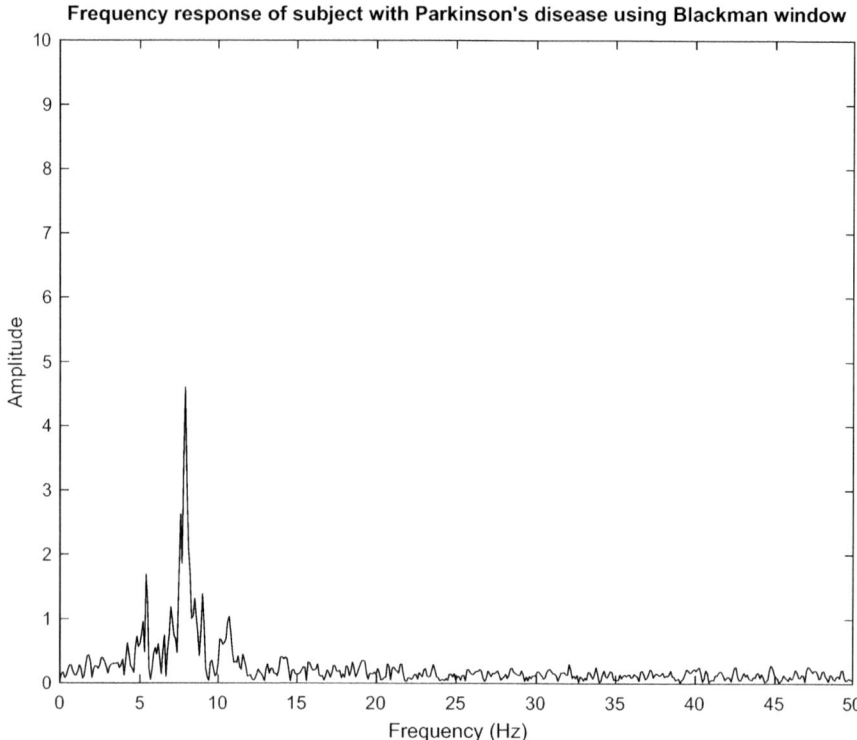

Fig. 7.8 Frequently domain using Blackman window of the acceleration magnitude signal for the subject with Parkinson's disease [7]

smartphone has access to the Internet for conveying data on the scale of a telecommunications footprint. However, the smartphone generally requires marginal cost, which may impact cost considerations. One notable advantage of the smartphone is its effectively ubiquitous presence and utility [1, 3–6, 9–11].

Not only do smartphones have the potential for objective quantification of movement disorder tremor with implications for titration of medication; the smartphone can provide quantified feedback with regard to deep brain stimulation tuning efficacy. Preliminary research, development, testing, and evaluation have been demonstrated for attaining considerable classification accuracy with regard to deep brain stimulation using "'On" and "'Off" status. The inertial signal data was consolidated into a feature set through software automation. These investigations have pertained to subjects with Parkinson's disease and Essential tremor [12–14].

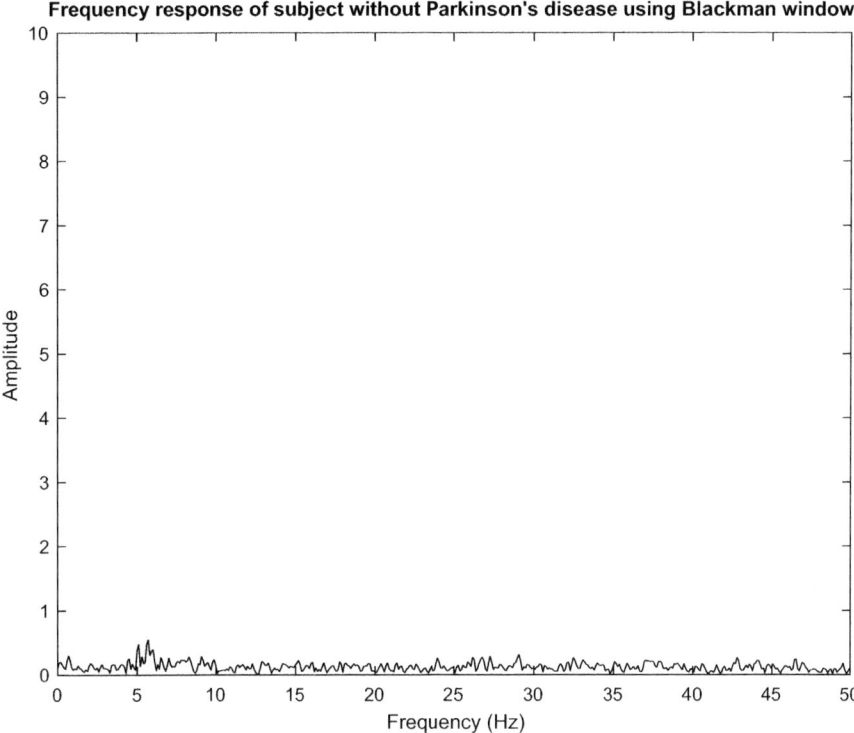

Fig. 7.9 Frequently domain using Blackman window of the acceleration magnitude signal for the subject without Parkinson's disease [7]

7.3   Smartphone for Quantification of Essential Tremor Regarding Deep Brain Stimulation in "On" and "Off" Mode

The successful first of its kind research investigation conducted by LeMoyne et al. during 2010 for the objective and quantified contrast of Parkinson's disease hand tremor to a subject without movement disorder demonstrated the nascent potential of wearable and wireless systems for the objective evaluation of movement disorder [7]. This technique can be extrapolated to facilitate the quantified evaluation of other movement disorders. For example, the smartphone as a wearable and wireless accelerometer platform can quantify the efficacy of sophisticated therapeutic intervention [15].

Another highly prevalent movement disorder is Essential tremor, which is comprised of both resting tremor and kinetic tremor. The kinetic tremor is observable during voluntary movement. Essential tremor is a progressive neurodegenerative disorder that is manifested predominantly about the extremities, and it can also influence the subject's voice and head [15].

Respective of scenarios for which traditional therapy, such as drug intervention, becomes intractable, deep brain stimulation offers a promising alternative for people with Essential tremor. For example, deep brain stimulation targeting the ventralis intermedius nucleus (VIM) of the thalamus offers an effective means for ameliorating movement disorder symptoms for Essential tremor. However, the programming of a deep brain stimulation system presents a challenging task, which is time-intensive. Furthermore, this endeavor requires the patient to travel to the clinician for a predetermined appointment. Also, the clinician must undergo additional specialized training to apply a subjectively interpreted algorithm to converge upon an optimal deep brain stimulation system parameter configuration [15].

A quantified methodology for measuring and recording tremor movement disorder for people with Essential tremor may enable rapid automated convergence to an optimal tuning configuration for deep brain stimulation. In addition, wearable and wireless systems equipped with inertial sensors can wirelessly transmit their signal data from a remote patient autonomous setting for pending analysis and post-processing by the respective clinical team. This amalgamation can better facilitate the maximal therapy strategy impact with minimal cost requirements. From a preliminary proof of concept from an engineering perspective, the application of a smartphone as a wearable and wireless accelerometer platform demonstrates this potential [15].

Preliminary objectives for the evolution and integration of wearable and wireless systems, such as a smartphone, seek to quantify tremor characteristics for a patient with Essential tremor using the smartphone as a functionally wireless accelerometer platform. The research applies analysis to the recorded acceleration signal for an Essential tremor patient treated with deep brain stimulation. The signal data acquired from the smartphone as a wireless accelerometer platform attempts to further develop the role of wearable and wireless systems for facilitating the programming of deep brain stimulation system parameter configurations. These objectives are demonstrated through applying the smartphone as a wearable and wireless accelerometer system for quantifying kinetic tremor for a subject with Essential tremor with deep brain stimulation status set to "On" and "Off" modes [15].

The methodology for the research endeavor involved one subject with Essential tremor. The subject's Essential tremor symptoms were under treatment through a deep brain stimulation system implanted with bilateral deep brain stimulation electrodes targeting the ventralis intermedius nucleus (VIM) nucleus of the thalamus. Experimental task for the subject was to reach for a lightweight object from a short distance on a table and then hold the object suspended. The characteristics of the task were quantified through a smartphone functioning as a wearable and wireless accelerometer platform secured about the dorsum of the hand through an elastic glove as illustrated in Fig. 7.10. The accelerometer signal data was transmitted through wireless connectivity to the Internet through email with the signal data attached to the email as a Comma-Separated-Value file. The experimental and post-processing resources were remotely situated transcontinentally from within the continental United States of America [15].

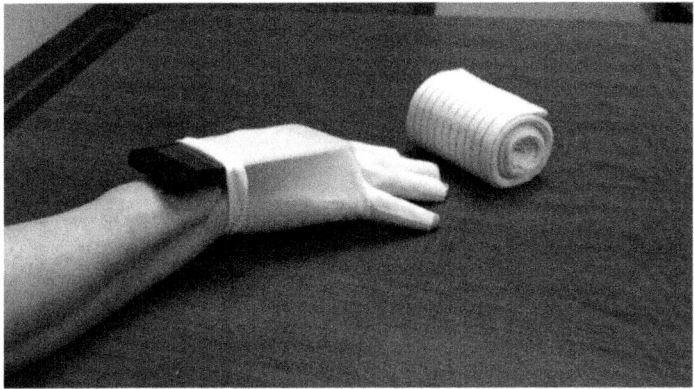

Fig. 7.10 Smartphone mounted to dorsum of the hand for subject with Essential tremor for reaching and grasping task regarding a lightweight object [15]

The following protocol was applied to conduct the experiment:

1. Mount the smartphone (iPhone) to the dorsum of the subject's hand using a latex glove.
2. Instruct the subject reach and grasp a lightweight object.
3. Activate the countdown sequence of the smartphone accelerometer recording.
4. Upon completion of the recording duration, email the accelerometer signal recording of the smartphone to an appropriate email.
5. Apply the protocol (1–4) for the following scenarios:

   (a) Deep brain stimulation "On" status
   (b) Deep brain stimulation "Off" status [15]

The results are illustrated in Figs. 7.11 and 7.12. These figures present the three orthogonal acceleration signals recorded by the smartphone and then conveyed by wireless transmission to the remotely situated post-processing resource. Figure 7.11 demonstrates smooth acceleration signals, while the deep brain stimulation system is activated to "On" status, which clearly illustrates the efficacy of the device for ameliorating movement disorder associated tremor. By contrast rhythmic oscillations of the accelerometer signals are clearly evidenced in Fig. 7.12, while the deep brain stimulation system is deactivated to "Off" mode [15].

The observable and quantified disparity of the acceleration signals for the deep brain stimulation system set to "On" and "Off" mode advocates the capacity to apply machine learning to successfully classify and distinguish between these two scenarios. In order to conduct machine learning classification, a series of appropriate numerical attributes should be selected, such as aspects of the temporal and frequency domains of the respective acceleration waveforms. This preliminary experimentation advocates the role of quantifying movement disorder tremor characteristics through a wearable and wireless system equipped with inertial sensors, such as a smartphone [15].

Fig. 7.11 Smartphone accelerometer signal during reaching and grasping task for Essential tremor subject; DBS set to "On" mode [15]

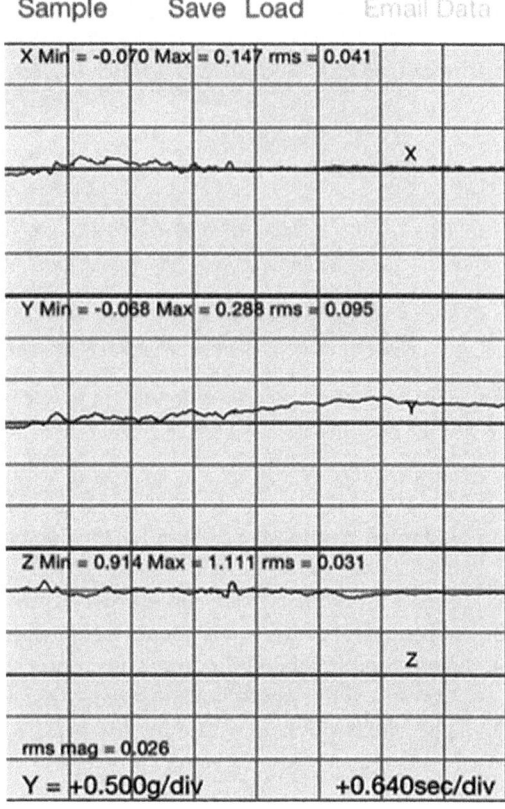

The inertial signal data for objective and quantified recording of movement disorder tremor can be wirelessly conveyed through connectivity to the Internet for later post-processing at a remotely situated location. The signal data can then be consolidated into a feature set for machine learning classification. For future endeavors the following machine learning algorithms warrant investigation:

- J48 decision tree
- Random forest
- Multilayer perceptron neural network
- Support vector machine
- K-nearest neighbors
- Logistic regression [15]

These research findings further advocate pathways toward the amalgamation of data science with movement disorder diagnosis and treatment. Historic databases using Cloud computing resources can objectively track the neurodegenerative status of an assortment of movement disorders. Furthermore, this novel technology could serve an instrumental role for deep brain stimulation programming. In essence, the

Fig. 7.12 Smartphone accelerometer signal during reaching and grasping task for Essential tremor subject; DBS set to "Off" mode [15]

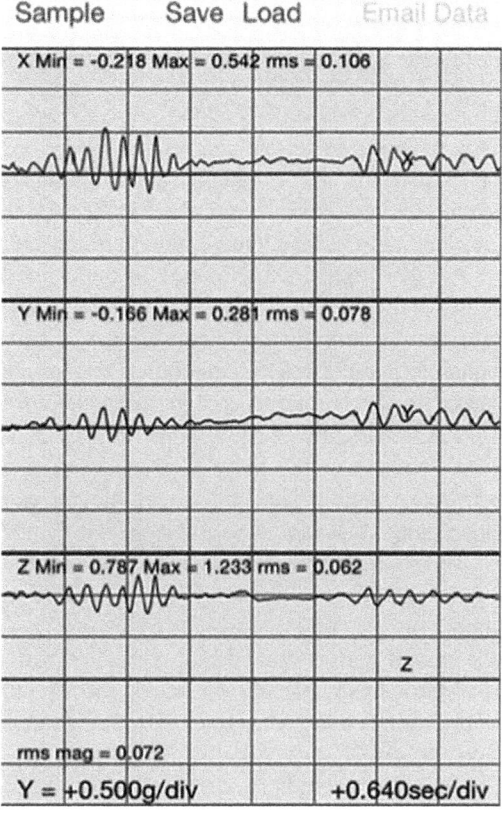

quantified feedback provided by the inertial signal of wearable and wireless systems could enable future closed-loop neurostimulation, such as deep brain stimulation devices, for optimal treatment of movement disorders [15].

7.4   Smartwatches, Bluetooth Wireless Connectivity, and Other Wearable and Wireless Systems for the Quantification of Movement Disorder Status

The use of wearable systems for monitoring health status is becoming more prevalent. These devices offer the potential to have transformative implications, such as advancing the quality of healthcare and alleviation of strain on limited medical economies, for people with movement disorders. In addition to logistical considerations for attending a clinical evaluation, the tendency for the movement disorder symptoms, such as for Parkinson's disease, to vary may result in the clinical evaluation that does not accurately represent the extent of the neurodegenerative disease [17].

The application of wearable systems can also potentially augment the clinician's acuity for the escalation to more progressively advanced therapy techniques. Subjects can have their movement disorder characteristics measured in the context of a remote setting through wearable sensors relative to the temporal bounds of clinical observation. This opportunity can present motor characteristics that may not be revealed to the clinician during a prescribed clinical period of monitoring. The application of remotely applied wearable sensors for recording movement disorder symptoms is advocated for the advance of diagnostic techniques to warrant the escalation of medical intervention to more sophisticated therapy strategies [18].

Another logical extrapolation of the capabilities of wearable and wireless systems for the inertial signal quantification of movement disorder tremor beyond the smartphone and portable media device is the smartwatch. The smartwatch is also equipped with wireless capabilities and an inertial sensor for quantifying tremor characteristics. For the configuration demonstrated by López-Blanco et al., the smartwatch locally conveys its inertial signal data to a smartphone using Bluetooth local wireless connectivity. The smartphone consolidates the accelerometer and gyroscope signal data into a text file for transmission to a computer for post-processing [19].

The research of López-Blanco et al. successfully demonstrated the ability of the smartwatch as a wearable and wireless inertial sensor system to quantify various modes of Essential tremor, such as resting, postural, and kinetic tremor. In addition, the inertial sensor signal data is correlated with clinical rating scales. Future extrapolation of this capability advocates the simplification and automation so that the application is more amenable to people without the skill set expertise of the clinical domain [19].

The smartwatch is an inherently wearable system that is equipped with an inertial sensor package. In particular the smartwatch is unique suited for continuous monitoring of fluctuating movement disorders, such as Essential tremor. The smartwatch applied in this research endeavor consists of an accelerometer and uses Bluetooth wireless to connect to a smartphone for wireless transmission to post-processing resources. The wearable and wireless device demonstrated the feasibility of the application for extended monitoring durations regarding people with Essential tremor with considerable correlation to clinical scale methodologies [20].

The primary advantage of the local Bluetooth wireless configuration is that the power requirements at the nodal level for the inertial sensor are greatly alleviated. Larger wearable and wireless systems, such as a smartphone, can then convey the inertial sensor signal trial data packet through wireless connectivity to the Internet for post-processing with a Cloud computing database anywhere in the world. Furthermore, multiple inertial sensor nodes with Bluetooth wireless connectivity can transmit their data to a locally situated and more powerful wearable and wireless system, such as a smartphone. Another alternative would be local wireless connectivity to a tablet [1, 16]. Figure 7.13 illustrates a conceptual perspective for the Bluetooth wireless transmission of signal data from the inertial sensor node level to a smartphone or tablet for subsequent wireless transmission to a Cloud computing resources for post-processing.

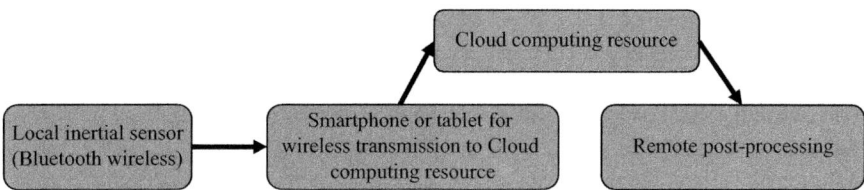

Fig. 7.13 Data signal pathway of a local inertial sensor to smartphone or tablet by Bluetooth wireless with subsequent wireless transmission to Cloud computing resource for post-processing

Rovini et al. advocate wearable and wireless systems for augmenting the diagnostic acuity of neurologists regarding the evaluation of Parkinson's disease. The device applies inertial sensors mounted about the fingers and the wrist, for which the signal data is conveyed to a local data logger through Bluetooth connectivity. The findings display a considerable degree of correlation between the wearable device and traditional clinical scales [21].

Kim et al. evaluate tremor severity due to Parkinson's disease through wearable and effectively wireless inertial sensors. The wearable application applies two sets of modules that mount to the wrist and finger. Each module is comprised of an inertial sensor system (accelerometer and gyroscope). The data stream from the finger module is transmitted by wire to the wrist module. From the wrist module data was conveyed through Bluetooth to a local computer for post-processing [22].

The inertial signal data was transformed to the frequency domain. Primary machine learning classification used a novel convolutional neural network architecture for distinguishing tremor severity. The concept proposed by Kim et al. provided considerable classification while transcending the performance of more traditional machine learning algorithms. Kim et al. advocate their methodology as potentially advancing acuity and possibly continuous monitoring for movement disorder symptoms in the context of daily life [22].

Further opportunity of wearable systems with potential for wireless application is demonstrated through the amalgamation of inertial sensors with the implementation of an orientation filter for advanced spatial awareness regarding movement disorder status. Van den Noort et al. developed a sensor-rich apparatus termed the PowerGlove for the quantification of motor characteristics of the hand for people with Parkinson's disease. The PowerGlove consists of 11 sensor units that were primarily comprised of accelerometers and gyroscopes situated to represent the spatial characteristics of the hand. The inertial signal data was applied to an extended Kalman filter for the development of a biomechanical model of the hand. A force sensor was also integrated into the PowerGlove for the goal of evaluating wrist rigidly. Although the current device is not wireless, the capacity to instill wireless capability is under consideration for future evolutions. For an assortment of hand movement scenarios, the PowerGlove successfully distinguished from a quantified perspective off-medication and on-medication conditions for subjects with Parkinson's disease [23].

Beyond the domain of technical suitability for the application of wearable and wireless systems for monitoring movement disorders status and eventual feedback for the optimal tuning configuration of deep brain stimulation exists the reaction of the user to this technology. There are multiple nontechnical challenges to the appropriate application of wearable and wireless systems in an autonomous environment. For example, the ease of mounting the device and impact toward the quality of personal activities influences the capacity of integration into daily life for extended monitoring. There is a psychosocial aspect regarding the level of personal comfort for wearing the device in a public setting. Another consideration is the level of technical sophistication required to sustainably operate the wearable and wireless system for an extended duration [24].

7.5   Network Centric Therapy

Wearable systems for the remote evaluation of Parkinson's disease are becoming increasingly integrated with the Internet of Things [25]. The emergence of the Internet of Things with the treatment of movement disorder symptoms, such as Parkinson's disease, has the potential to have a transformative influence respective of the global healthcare domain especially pertaining to diagnostics and treatment [26]. These technology trends establish precedence for development of Network Centric Therapy for the optimal treatment of movement disorders, such as Parkinson's disease and Essential tremor, which is inclusive of data science strategies, which has also been advocated for domains, such as rehabilitation for hemiparesis [1, 27].

The environment of Network Centric Therapy for the treatment of neurodegenerative movement disorders, such as Parkinson's disease and Essential tremor, forecasts the acquired inertial sensor signal data being stored in a Cloud computing database. The signal data is derived from the application of wearable and wireless systems. With the location of the experimental site remote to the post-processing resources, specialized clinical intervention can occur from anywhere in the world, which is the essence of Network Centric Therapy. Network Centric Therapy can facilitate feedback interactivity for the optimized proactive intervention strategy of neurodegenerative movement disorders, such as Parkinson's disease and Essential tremor. Preliminary conceptualization of Network Centric Therapy is illuminated by Fig. 7.14.

Machine learning constitutes a critical synergy with Network Center Therapy. From a post-processing perspective, machine learning classification accuracy can distinguish and essentially diagnose the health status for a patient with a neurodegenerative movement disorder, such as Parkinson's disease and Essential tremor. A vantage of the role of machine learning for classifying movement disorder symptoms with respect to deep brain stimulation is provided in Chap. 8 "Role of machine learning for classification of movement disorder and deep brain stimulation status."

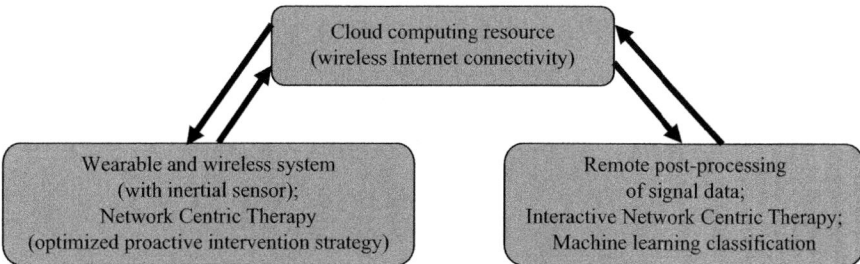

Fig. 7.14 Schematic process of Network Centric Therapy configuration with wearable and wireless systems achieving Internet connectivity to Cloud computing databases, the bi-directional arrows imply feedback between the subject and post-processing resources

7.6   Conclusion

Preliminary manifestations of wearable and wireless systems with access to the Internet have been successfully demonstrated through the research, development, test, and evaluation of smartphones for the quantification of Parkinson's disease hand tremor. The device was readily mounted to the dorsum of the hand through a glove. Analysis of the acceleration waveform for quantified parameters displayed statistical significance between a patient with Parkinson's disease and a subject without Parkinson's disease. Preliminary origins of Network Centric Therapy were elucidated, since the experimentation site and post-processing resources were situated on effectively opposite sides of the continental United States. Other proposed wearable and wireless systems involve the consideration of portable media devices.

Wearable and wireless systems have been further extrapolated for the role of a feedback methodology regarding the efficacy of deep brain stimulation. For a subject with Essential tremor, the inertial sensor signals derived from the smartphone as a wearable and wireless systems identified notable disparity respective of the deep brain stimulation system set to "On" and "Off" status. Future extensions recommend machine learning as an advanced diagnostic technique. Eventually wearable and wireless systems are envisioned to provide feedback for closed-loop acquisition of optimal parameter configurations.

Advanced utility of wearable and wireless systems is envisioned with the role of Bluetooth wireless for local connectivity. This scheme seizes the opportunity presented by dual wireless connectivity. An inertial sensor node achieves local wireless connectivity to a smartphone or tablet for pending transmission of the data package through more global connectivity to the Internet for Cloud computing database storage for remote post-processing. For example, the inertial sensor signal of a smartwatch can be locally and wirelessly transmitted to a smartphone. This approach enabled further minimization of the inertial sensor node. These observed trends advocate the imminent presence of Network Centric Therapy for the patient-specific optimized treatment regarding neurodegenerative movement disorders, such as Parkinson's disease and Essential tremor.

References

1. LeMoyne R, Mastroianni T (2018) Wearable and wireless systems for healthcare I: gait and reflex response quantification. Springer, Singapore
2. LeMoyne R, Coroian C, Cozza M, Opalinski P, Mastroianni T, Grundfest W (2009) The merits of artificial proprioception, with applications in biofeedback gait rehabilitation concepts and movement disorder characterization. In: Biomedical engineering. InTech, Vienna, pp 165–198
3. LeMoyne R, Mastroianni T (2017) Smartphone and portable media device: a novel pathway toward the diagnostic characterization of human movement. In: Smartphones from an applied research perspective. InTech, Rijeka, Croatia, pp 1–24
4. LeMoyne R, Mastroianni T (2017) Wearable and wireless gait analysis platforms: smartphones and portable media devices. In: Wireless MEMS networks and applications. Elsevier, New York, pp 129–152
5. LeMoyne R, Mastroianni T (2016) Telemedicine perspectives for wearable and wireless applications serving the domain of neurorehabilitation and movement disorder treatment. In: Telemedicine, SMGroup, Dover, Delaware, pp 1–10
6. LeMoyne R, Mastroianni T (2015) Use of smartphones and portable media devices for quantifying human movement characteristics of gait, tendon reflex response, and Parkinson's disease hand tremor. In: Mobile health technologies, methods and protocols. Springer, New York, pp 335–358
7. LeMoyne R, Mastroianni T, Cozza M, Coroian C, Grundfest W (2010) Implementation of an iPhone for characterizing Parkinson's disease tremor through a wireless accelerometer application. In: 32nd annual international conference of the IEEE, Engineering in Medicine and Biology Society (EMBS), pp 4954–4958
8. Kostikis N, Hristu-Varsakelis D, Arnaoutoglou M, Kotsavasiloglou C, Baloyiannis S (2011) Towards remote evaluation of movement disorders via smartphones. In: 33rd annual international conference of the IEEE, Engineering in Medicine and Biology Society (EMBS), pp 5240–5243
9. LeMoyne R, Mastroianni T, Grundfest W (2012) Quantified reflex strategy using an iPod as a wireless accelerometer application. In: 34th annual international conference of the IEEE, Engineering in Medicine and Biology Society (EMBS), pp 2476–2479
10. LeMoyne R, Mastroianni T, Grundfest W, Nishikawa K (2013) Implementation of an iPhone wireless accelerometer application for the quantification of reflex response. In: 35th annual international conference of the IEEE, Engineering in Medicine and Biology Society (EMBS), pp. 4658–4661
11. LeMoyne R, Mastroianni T (2017) Implementation of a smartphone wireless gyroscope platform with machine learning for classifying disparity of a hemiplegic patellar tendon reflex pair. J Mech Med Biol 17(6):1750083
12. LeMoyne R, Tomycz N, Mastroianni T, McCandless C, Cozza M, Peduto D (2015) Implementation of a smartphone wireless accelerometer platform for establishing deep brain stimulation treatment efficacy of essential tremor with machine learning. In: 37th annual international conference of the IEEE, Engineering in Medicine and Biology Society (EMBS), pp 6772–6775
13. LeMoyne R, Mastroianni T, Tomycz N, Whiting D, Oh M, McCandless C, Currivan C, Peduto D (2017) Implementation of a multilayer perceptron neural network for classifying deep brain stimulation in 'On' and 'Off' modes through a smartphone representing a wearable and wireless sensor application. In: 47th Society for Neuroscience annual meeting (featured in Hot Topics; top 1% of abstracts)
14. LeMoyne R, Mastroianni T, McCandless C, Currivan C, Whiting D, Tomycz N (2018) Implementation of a smartphone as a wearable and wireless accelerometer and gyroscope platform for ascertaining deep brain stimulation treatment efficacy of Parkinson's disease through machine learning classification. Adv Park Dis 7(2):19–30

15. LeMoyne R, Mastroianni T, Tomycz N, Whiting D, McCandless C, Peduto D, Cozza M (2015) I-Phone wireless accelerometer quantification of extremity tremor in essential tremor patient undergoing activated and inactivated deep brain stimulation. In: International Neuromodulation Society's 12th World Congress

16. LeMoyne R, Mastroianni T (2018) Bluetooth inertial sensors for gait and reflex response quantification with perspectives regarding cloud computing and the Internet of Things. In: Wearable and wireless systems for healthcare I: gait and reflex response quantification. Springer, Singapore, pp 95–103

17. Heldman DA, Harris DA, Felong T, Andrzejewski KL, Dorsey ER, Giuffrida JP, Goldberg B, Burack MA (2017) Telehealth management of Parkinson's disease using wearable sensors: an exploratory study. Digit Biomark 1(1):43–51

18. Heldman DA, Giuffrida JP, Cubo E (2016) Wearable sensors for advanced therapy referral in Parkinson's disease. J Park Dis 6(3):631–638

19. López-Blanco R, Velasco MA, Méndez-Guerrero A, Romero JP, del Castillo MD, Serrano JI, Benito-León J, Bermejo-Pareja F, Rocon E (2018) Essential tremor quantification based on the combined use of a smartphone and a smartwatch: the NetMD study. J Neurosci Methods 303:95–102

20. Zheng X, Vieira Campos A, Ordieres-Meré J, Balseiro J, Labrador Marcos S, Aladro Y (2017) Continuous monitoring of essential tremor using a portable system based on smartwatch. Front Neurol 8:96

21. Rovini E, Esposito D, Maremmani C, Bongioanni P, Cavallo F (2014) Using wearable sensor systems for objective assessment of Parkinson's disease. In: 20th IMEKO TC4 international symposium and 18th international workshop on ADC modelling and testing, pp 862–867

22. Kim HB, Lee WW, Kim A, Lee HJ, Park HY, Jeon HS, Kim SK, Jeon B, Park KS (2018) Wrist sensor-based tremor severity quantification in Parkinson's disease using convolutional neural network. Comput Biol Med 95:140–146

23. van den Noort JC, Verhagen R, van Dijk KJ, Veltink PH, Vos MC, de Bie RM, Bour LJ, Heida CT (2017) Quantification of hand motor symptoms in Parkinson's disease: a proof-of-principle study using inertial and force sensors. Ann Biomed Eng 45(10):2423–2436

24. Johansson D, Malmgren K, Murphy MA (2018) Wearable sensors for clinical applications in epilepsy, Parkinson's disease, and stroke: a mixed-methods systematic review. J Neurol 265(8):1740–1752

25. Rovini E, Maremmani C, Cavallo F (2018) Automated systems based on wearable sensors for the management of Parkinson's disease at home: a systematic review. Telemed E-Health (Epub ahead of print)

26. Pasluosta CF, Gassner H, Winkler J, Klucken J, Eskofier BM (2015) An emerging era in the management of Parkinson's disease: wearable technologies and the Internet of Things. IEEE J Biomed Health Inform 19(6):1873–1881

27. LeMoyne R, Mastroianni T (2018) Future perspective of network centric therapy. In: Wearable and wireless systems for healthcare I: gait and reflex response quantification. Springer, Singapore, pp 133–134

Chapter 8
Role of Machine Learning for Classification of Movement Disorder and Deep Brain Stimulation Status

Abstract Recently, machine learning has augmented the capability of the amalgamation of wearable and wireless systems for deep brain stimulation systems. Machine learning platforms have been applied to attain considerable classification accuracy for distinguishing between deep brain stimulation set to "On" and "Off" modes for Essential tremor and Parkinson's disease. Other movement disorders, such as hemiplegic affected and unaffected limb pairs, have been successfully differentiated through machine learning classification. Central to these machine learning endeavors has been the application of wearable and wireless systems using inertial sensors, such as the accelerometer and gyroscope, to consolidate signal data into feature sets for machine learning classification. An assortment of prevalent machine learning platforms is discussed, such as J48 decision tree, K-nearest neighbor, logistic regression, support vector machine, multilayer perceptron neural network, and random forest. Machine learning is envisioned to serve an instrumental role for the objective of achieving closed-loop optimization of deep brain stimulation parameter configurations. In essence, machine learning is envisioned to function as a predominant role for the post-processing perspective of Network Centric Therapy.

Keywords Machine learning · Waikato Environment for Knowledge Analysis (WEKA) · J48 decision tree · K-nearest neighbors · Logistic regression · Support vector machine · Multilayer perceptron neural network · Random forest · Wireless accelerometer · Wireless gyroscope · Smartphone · Portable media device · Essential tremor · Parkinson's disease · Movement disorder · Hand tremor

8.1 Introduction

The amalgamation of machine learning with the wearable and wireless inertial sensors for the objective quantification of movement disorder status offers the potential to considerably transcend the utility of conventional diagnostic techniques. Machine learning classification through the application of wearable and wireless systems to

© Springer Nature Singapore Pte Ltd. 2019   99
R. LeMoyne et al., *Wearable and Wireless Systems for Healthcare II*,
Smart Sensors, Measurement and Instrumentation 31,
https://doi.org/10.1007/978-981-13-5808-1_8

establish an inertial signal derived feature set offers the opportunity to distinguish the efficacy of disparate medical therapy interventions and even deep brain stimulation parameter configurations [1–5]. Recently, machine learning has been proposed and applied for deep brain stimulation using feedback from the inertial sensor signal provided by a wearable and wireless system [6–9]. Considerable classification accuracy was attained for the distinction between deep brain stimulation scenarios, such as "On" and "Off" configurations for people with neurodegenerative movement disorders, such as Essential tremor and Parkinson's disease [6–8]. The integral application of machine learning for deep brain stimulation using wearable and wireless systems for objectively quantified accelerometer and gyroscope signal feedback is envisioned to enable the pathway to closed-loop deep brain stimulation systems capable of acquiring automated and optimized parameter configurations.

The Waikato Environment for Knowledge Analysis (WEKA) offers an assortment of machine learning platforms. In particular, six frequently utilized machine learning techniques are J48 decision tree, K-nearest neighbors, logistic regression, support vector machine, multilayer perceptron neural network, and random forest [10–12]. The proper selection of a machine learning algorithm is uniquely suited to the context of the classification endeavor under consideration [13]. Therefore, a more in-depth consideration of the inherent nature of these machine learning algorithms is addressed.

8.2   Waikato Environment for Knowledge Analysis (WEKA) for Machine Learning Classification of Movement Disorder Ameliorated Through Deep Brain Stimulation Using Wearable and Wireless Systems for Quantified Feedback

A prevalent machine learning platform is the Waikato Environment for Knowledge Analysis (WEKA). WEKA enables the scientific community to apply a considerable assortment of machine learning classification algorithms. In particular, WEKA incorporates a highly operable graphic user interface that facilitates the ability to conduct machine learning classification endeavors [10–12]. The machine learning classification platforms offered by WEKA have been successfully applied with regard to movement disorder scenarios for distinguishing various deep brain stimulation configurations, such as "On" and "Off" status [6–8]. Figure 8.1 presents a representative graphic user interface for WEKA pertaining to the machine learning classification of deep brain stimulation "On" and "Off" status.

Machine learning classification requires the organization of a feature set. The feature set is comprised of attributes and classes. The classes represent the types of data to be distinguished, such as disparate neurological movement disorder scenarios, through machine learning. The attributes pertain to numeric values that appropriately describe the respective classes under consideration [10–12]. Post-processing

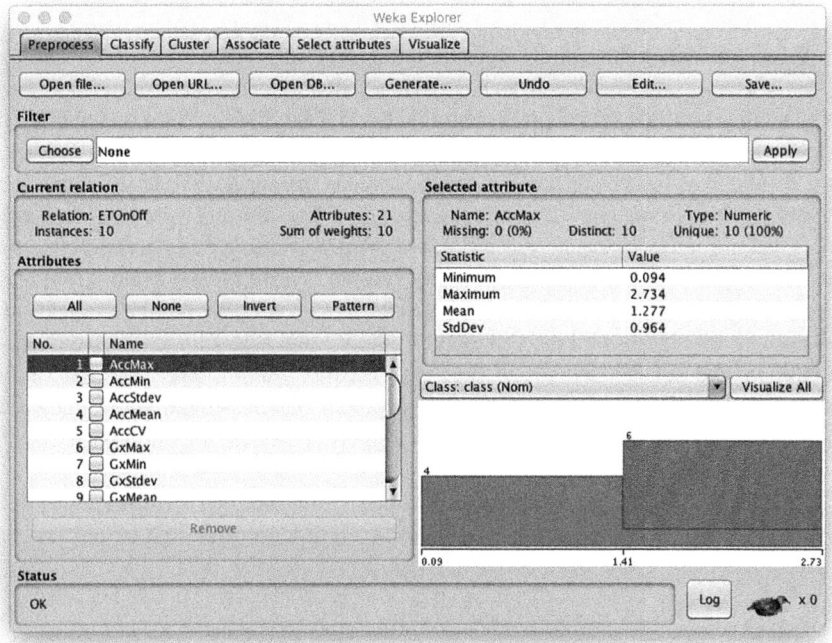

Fig. 8.1 A representative graphic user interface for WEKA regarding classification to distinguish between deep brain stimulation "On" and "Off" status for Essential tremor [8, 10–12]

techniques regarding data acquired from wearable and wireless systems, such as the smartphone and portable media device, have been applied to consolidate feature sets for successful machine learning classification of disparate scenarios of neurological movement disorders [6–8, 14–18].

The feature set is consolidated through the post-processing of the available iner-tial signal data, such as the accelerometer and/or gyroscope. The selected numeric attributes of the feature set are composed at the discretion of the research team conducting the machine learning endeavor. Upon observation of the available signal data, the proper numeric attributes can be selected, such as descriptive statistics that characterize the signal data and aspects of the frequency domain. Consolidation of the feature set is standardly conducted through the application of an automation software, which especially provides consistency. The product of the software auto-mation yields an Attribute-Relation File Format (ARFF) file comprised of attributes and their respective classes [6–8, 14–19].

The most significant quantified result that WEKA provides is the classification accuracy pertaining to the feature set. The classification accuracy provides the per-centage of correctly classified instances to the total number of instances. Another resultant product is the confusion matrix, which further defines the nature of the classification accuracy. The confusion matrix addresses the number of instances that were correctly and incorrectly classified through a matrix approach [10–12].

There are machine learning algorithms featured in WEKA that offer visualization of the machine learning strategy. Two prevalent strategies are the J48 decision tree and the multilayer perceptron neural network. The J48 decision tree applied by WEKA provides the decision tree that enables the optimal classification accuracy. An advantage of the visualized J48 decision tree is the capacity to reveal the logical structure for deriving the classification accuracy and the most predominant attributes that influence the decision tree [10–12]. A representative J48 decision tree that pertains to the classification of Essential tremor regarding deep brain stimulation therapy in "On" and "Off" mode is presented in Fig. 8.2. The WEKA classify tab window is displayed in Fig. 8.3, which shows the selection of tenfold cross-validation, the selected classifier, and the classifier output. The classifier output aspect of the WEKA classify tab window provides the classification accuracy with the percentage and number of correctly classified instances in the top right and the confusion matrix at the lower left [8].

Another visualized machine learning strategy provided by WEKA is the multi-layer perceptron neural network. The input layer, hidden layer, and output layer of

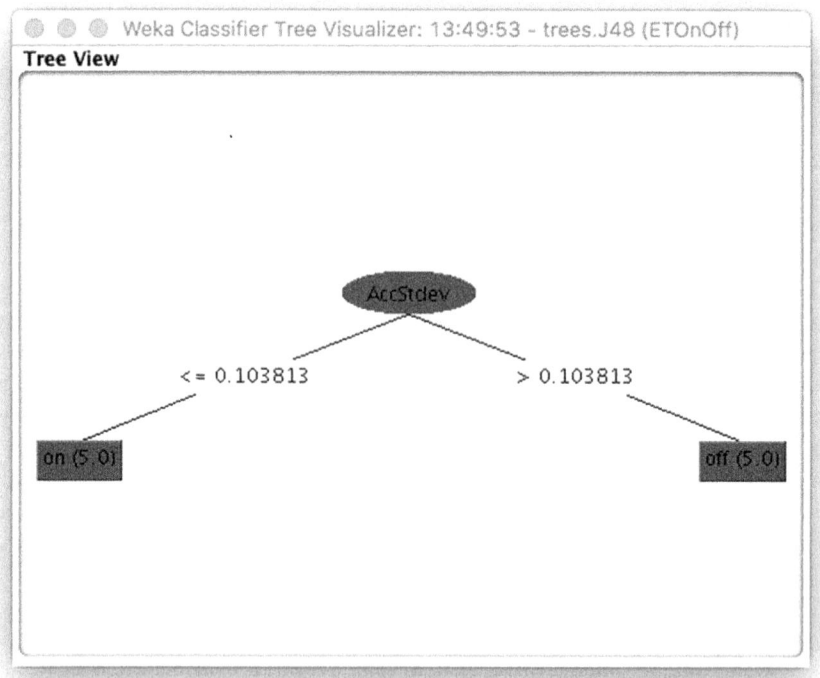

Fig. 8.2 J48 decision tree for distinguishing between deep brain stimulation "On" and "Off" status for Essential tremor [8]

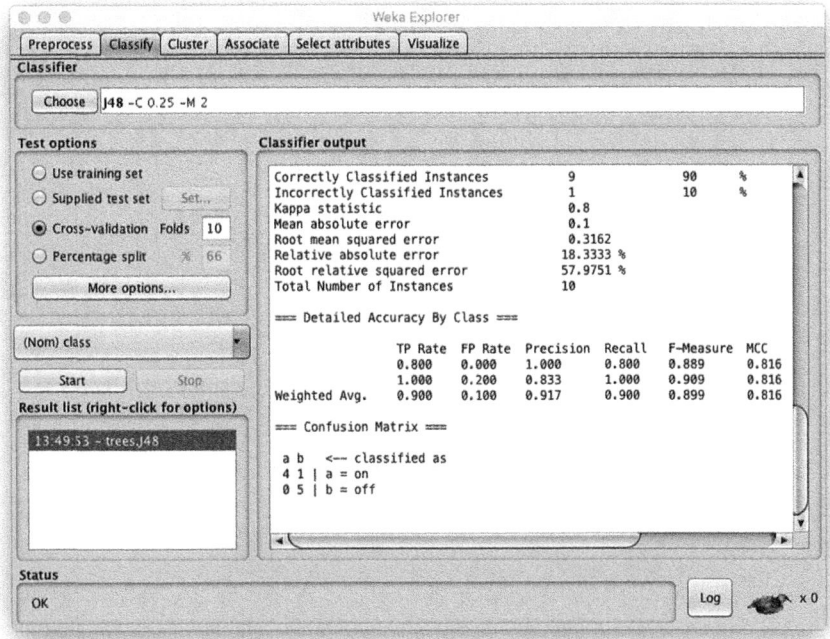

Fig. 8.3 WEKA classifier window for J48 decision tree [8]

the multilayer perceptron neural network are visually presented. The attributes of the feature set are represented by the inner layer. The output layer consists of the classes for the feature set, for which the hidden layers interconnect the input layer and output layer [7, 8, 16, 17]. Figure 8.4 presents a multilayer perceptron neural network for the machine learning classification of Essential tremor regarding deep brain stimulation therapy in "On" and "Off" mode. The WEKA classify tab window is presented in Fig. 8.5, which selects tenfold cross-validation, the chosen classifier, and the classifier output. The classifier output aspect of the WEKA classify tab window indicates the classification accuracy defined by the percentage and number of correctly classified instances near the top right and the confusion matrix at the lower left [8].

WEKA provides a scientific research team a numerous assortment of machine learning classification strategies [10–12]. In light of the multitude of machine classification algorithms available, the proper algorithm is highly dependent on the intrinsic characteristics of the feature set [13, 14]. Notably there are six machine learning classification techniques that warrant additional detail:

1. J48 decision tree
2. K-nearest neighbors
3. Logistic regression
4. Support vector machine

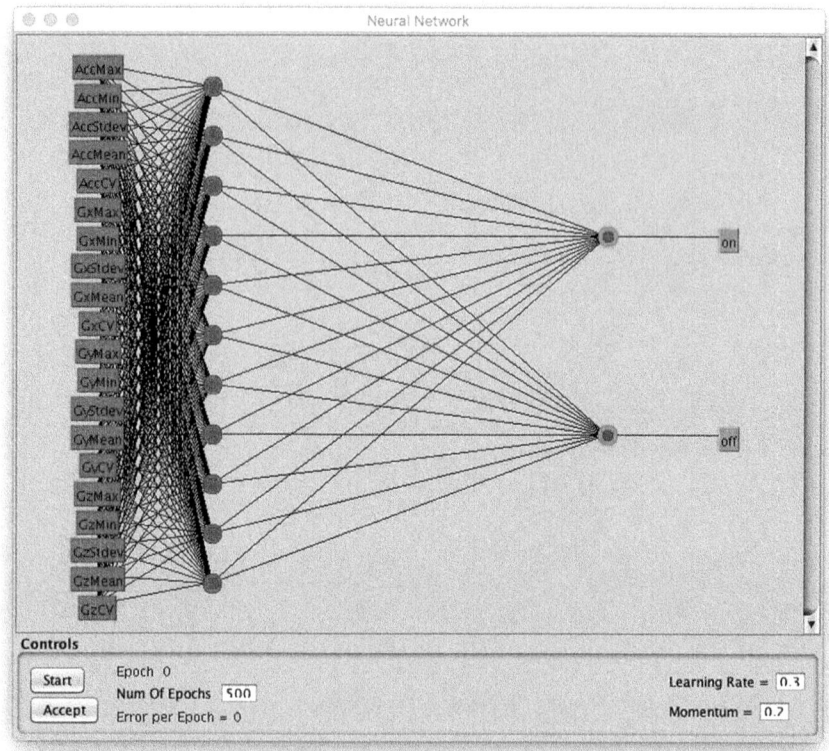

Fig. 8.4 Multilayer perceptron neural network consisting of an input layer (feature set attributes), hidden layer, output layer (classes) for distinguishing between deep brain stimulation "On" and "Off" status for Essential tremor [8]

5. Multilayer perceptron neural network
6. Random forest [10–12]

These algorithms for machine learning classification have been uniquely applied for the classification of neurological movement disorder scenarios through the objective quantification of the inertial signal enabled through the wearable and wireless systems [6–8, 14–18].

8.2.1   J48 Decision Tree

The J48 decision tree provided by WEKA utilizes the C4.5 decision tree algorithm. A unique advantage of the WEKA J48 decision tree is the capability to visualize the relevant decision tree. The benefit of a graphical representation of the decision tree is the elucidation of the predominant attributes used to define the branches of the

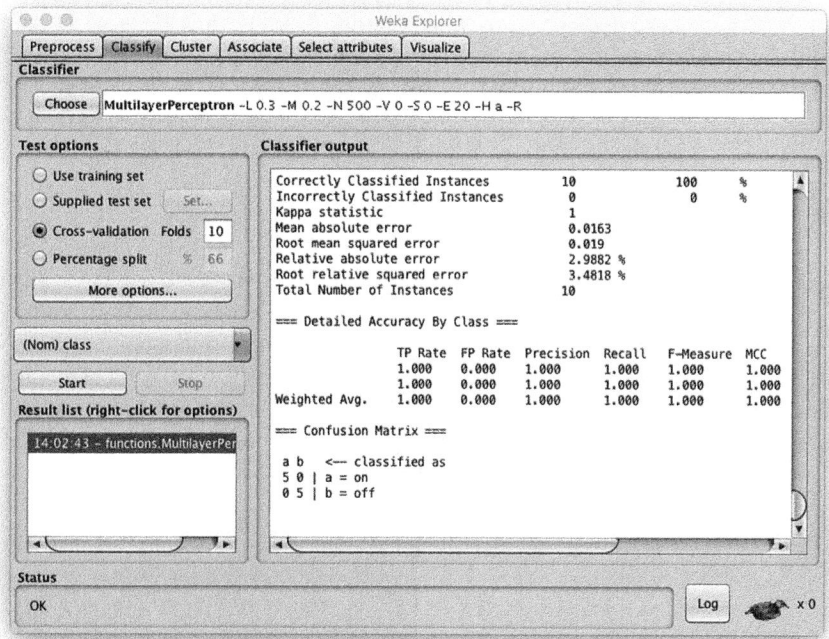

Fig. 8.5 Weka classifier window for multilayer perceptron neural network [8]

decision tree [10–12]. The J48 decision tree has achieved considerable classification accuracy for the differentiation of spastic movement disorder pertaining to a hemiplegic reflex pair using the gyroscope signal of a smartphone representing a wearable and wireless system [18, 20].

The J48 algorithm optimizes the decision tree through the application of information theory. This algorithm applies a top-down strategy through a recursive divide and conquer technique, which maximizes information gain through the quantification of entropy. Information is measured in bits. The process derives a decision tree emanated from a root node with a sequence of branches that evaluate a specific attribute in terms of a Boolean logical condition to eventually lead to an established class [10–12].

8.2.2 K-Nearest Neighbors

The K-nearest neighbors machine learning algorithm represents a robust technique that has its origins dating back to the 1950s, which has been advocated by statisticians. This approach applies an instance-based learning strategy. The algorithm collects piecewise linear decision boundaries. The objective of K-nearest neighbors is to determine the most similar training set by comparison to the new instance [10–12]. The K-nearest neighbors machine learning strategy has been successfully applied toward distinguishing movement disorder, such as spasticity, regarding a hemiplegic tendon reflex pair with considerable classification accuracy [18].

8.2.3   Logistic Regression

Logistic regression is another machine learning algorithm made readily available through WEKA. Foundational to the concept of logistic regression is the logit transform. The logit transform is numerically represented from a computational perspective by application of a sigmoidal function [10–12]. Logistic regression has attained considerable classification accuracy for the distinction of a hemiplegic ankle/foot complex pair with one side influenced by spastic movement disorder. A benefit for the application of logistic regression is the ability to address the broad range regarding the change in distribution for the pathology under consideration for machine learning classification [13].

8.2.4   Support Vector Machine

The support vector machine learning first represents the feature set into a hyperspace by a mathematical technique. Through the application of a kernel, the hyperspace is consolidated to a hyperplane. Machine learning classification is achieved through differentiating the classes that define the feature set through a support vector [10–12, 21].

The support vector machine has been successfully applied to the domain of classification of movement disorder scenarios using the wearable and wireless system to provide inertial sensor signal data to establish the relevant feature set. Regarding spastic movement disorder associated with hemiparesis, the support vector machine achieved considerable classification accuracy to distinguish between the affected and unaffected leg with respect to both the patellar tendon reflex and ankle/foot complex [14, 15]. Using the smartphone as a wearable and wireless accelerometer platform to provide the basis for the feature set, the support vector machine achieved considerable classification accuracy to distinguish between deep brain stimulation set to "On" and "Off" status for the reaching and grasping of a lightweight object [6].

8.2.5   Multilayer Perceptron Neural Network

The multilayer perceptron neural network presents a biomimetic form of machine learnings, since it provides a computational basis for the foundational perceptivity of the brain at the neuronal level [10–12, 22]. The multilayer perceptron neural network has been successfully applied in the context of wearable and wireless systems with regard to therapy interventions and spastic movement disorder [17, 23–26]. Signal data from the smartphone as a wearable and wireless inertial sensor platform, such as the accelerometer and gyroscope, has been consolidated into a feature set for machine learning classification with the multilayer perceptron neural network. The multilayer perceptron neural network has attained considerable classification accuracy for distinguishing between the deep brain stimulation system in "On" and "Off" status for both Essential tremor and Parkinson's disease [7, 8, 27].

8.2.6 Random Forest

The random forest machine learning technique incorporates the foundation of the decision tree approach. A series of randomized decision trees are generated by the machine learning algorithm. The randomization process achieves a diverse perspective regarding the selection of the most optimal decision tree. This technique applied regarding the random forest provides an improved decision tree [10–12].

8.2.7 Attribute-Relation File Format (ARFF)

The Attribute-Relation File Format (ARFF) represents the feature set for machine learning classification using WEKA. The ARFF constitutes a consolidation of experimental data that is synthesized into numerical attributes. The numeric attributes of the feature set are determined at the discretion of the research team conducting the machine learning classification endeavor. The objective is to select the attributes that most appropriately represent the characteristics of the experiment [10–12]. For example, while addressing the inertial signal for a wearable and wireless system to quantify hand tremor status for either Parkinson's disease or Essential tremor for various scenarios involving deep brain stimulation, the descriptive statistics of the accelerometer or gyroscope signal would be suitable to represent the numeric attributes that compose the feature set [6–8].

The ARFF for WEKA can be represented so that they can be manually manipulated through other software platforms, such as Excel. The ARFF can be saved and loaded in Comma-Separated-Value (CSV) format. In CSV form the research team can modify targeted numeric attribute parameters to the ARFF at their discretion [10–12].

Another technique is to apply software automation to generate the ARFF. Two programming languages that have been demonstrated for the automated development of the ARFF are Matlab and Python. Wearable and wireless systems that incorporate the inertial sensor platform of the smartphone and portable media device output their quantified signal data as CSV for each experimental trial. The preliminary state of the ARRF file consists of defining the characteristics of the attributes and the classes that will be distinguished through machine learning. The automation

Fig. 8.6 Algorithm process for deriving a feature set through machine learning classification to differentiate between distinct movement disorder scenarios

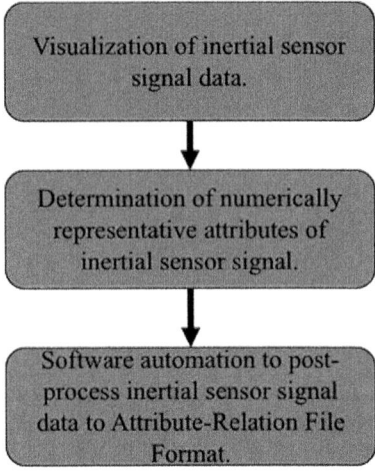

programs first extract the relevant data from the CSV file consisting of the signal data. Then numerical methods are applied to post-process the data to the desired numeric attributes representing the feature set. The numeric attribute data for each trial is sequentially amended to the ARFF until all instances for the experimental data are included [3, 6–8, 13, 14, 16–18, 24–26]. Figure 8.6 presents a representative algorithm process for post-processing inertial sensor signal data from a wearable and wireless system into a feature set for identifying disparity of movement disorder scenarios from a quantified perspective.

8.3   The Role of Machine Learning for Network Centric Therapy

Machine learning is anticipated to represent an inherent feature for the evolving development of Network Centric Therapy. Using the wearable and wireless system to establish a feature set for machine learning classification as substantial improvement for the treatment of neurodegenerative movement disorders, such as Parkinson's disease and Essential tremor, is envisioned. Once a machine learning algorithm has been successfully applied for an appropriately representative patient sample, the technique can then be provided to far larger patient samples for their respective diagnosis and therapy intervention.

Beyond the scope of conventional treatment intervention is the domain of deep brain stimulation. The response to the considerable array of deep brain stimulation parameter configurations can be quantified by wearable and wireless systems through their inertial sensor signal data. However, the time constraint for even an expert clinician to acquire the response of all available deep brain stimulation parameter configurations is intuitively a dauntingly laborious task, which may be in fact impractical. An automation strategy that implements machine learning may represent a breakthrough development for the acquisition of closed-loop optimal deep brain stimulation.

As previously discussed an assortment of machine learning algorithms are available for the objective of serving a critical role for the achievement of an automated closed-loop parameter configuration optimization strategy for deep brain stimulation. Depending on the computational complexity of the machine learning algorithm, the classifier could either be implemented through the smartphone or the Cloud computing environment. In essence, the machine learning classification algorithm could adapt to the patient-specific unique characteristics, such that the closed-loop optimization of the deep brain stimulation system configuration parameters could be effectively achieved on a real-time basis.

Chapter 9 "Assessment of machine learning classification strategies for the differentiation of deep brain stimulation 'On' and 'Off' status for Parkinson's disease" provides preliminary insight as to the proposed amalgamation of machine learning for wearable and wireless systems applied for deep brain stimulation. An assortment of machine learning algorithms is evaluated for the ability to achieve appreciable classification accuracy. The two classes to be distinguished are with respect to setting the deep brain stimulation system to "On" and "Off" mode. The subject for this engineering proof of concept perspective was diagnosed with Parkinson's disease. However, other neurodegenerative movement disorders, such as Essential tremor, are suitable to the general methodology with minor modification of the experimental protocol. Rather than utilize a Cloud computing environment for storage of the inertial sensor signal data, the data is stored in a provisional email resource. With the data downloaded to a local personal computer for post-processing, a software program automates the consolidation of the inertial sensor signal data to a feature set for post-processing. The machine learning classification algorithms are subsequently compared and contrasted in terms of classification accuracy and computational robustness for feasibility with respect to further evolution toward Network Centric Therapy.

8.4 Conclusion

The integration of machine learning enables substantial opportunity for the use of wearable and wireless systems, such as a smartphone, with operation of deep brain stimulation for the treatment of progressive neurodegenerative movement disorders, such as Parkinson's disease and Essential tremor. The Waikato Environment for Knowledge Analysis (WEKA) is featured as a robust machine learning program with an assortment of machine learning algorithms. Emphasis is drawn to six machine learning techniques: J48 decision tree, K-nearest neighbor, logistic regression, support vector machine, multilayer perceptron neural network, and random forest. The inertial sensor signal, such as an accelerometer and gyroscope, acquired by the wearable and wireless system provides a basis for establishing the feature set for machine learning classification. A software automation program is generally applied for the consolidation of the signal data into a feature set. For WEKA the feature set is presented as an Attribute-Relation File Format (ARFF). Considerable machine learning classification accuracy has been attained to distinguish between disparate scenarios regarding movement disorders. In particular considerable machine learning classification accuracy has been achieved to differentiate between deep brain stimulation "On" and "Off" status for neurodegenerative movement disorders, such as Parkinson's disease and Essential tremor. These achievements advocate and forecast the presence of machine learning for augmenting the capability to optimize a deep brain stimulation parameter configuration potentially in real-time. For Network Centric Therapy, machine learning is envisioned to serve a major role regarding the post-processing of the signal data acquired by the wearable and wireless systems as quantified feedback for the treatment intervention provided by deep brain stimulation for neurodegenerative movement disorders, such as Parkinson's disease and Essential tremor.

References

1. LeMoyne R, Mastroianni T (2015) Use of smartphones and portable media devices for quantifying human movement characteristics of gait, tendon reflex response, and Parkinson's disease hand tremor. In: Mobile health technologies, methods and protocols. Springer, New York, pp 335–358
2. LeMoyne R, Mastroianni T (2017) Wearable and wireless gait analysis platforms: smartphones and portable media devices. In: Wireless MEMS networks and applications. Elsevier, New York, pp 129–152
3. LeMoyne R, Mastroianni T (2016) Telemedicine perspectives for wearable and wireless applications serving the domain of neurorehabilitation and movement disorder treatment. In: Telemedicine, SMGroup, Dover, Delaware, pp 1–10
4. LeMoyne R, Coroian C, Cozza M, Opalinski P, Mastroianni T, Grundfest W (2009) The merits of artificial proprioception, with applications in biofeedback gait rehabilitation concepts and movement disorder characterization. In: Biomedical engineering. InTech, Vienna, pp 165–198
5. LeMoyne R, Mastroianni T (2017) Smartphone and portable media device: a novel pathway toward the diagnostic characterization of human movement. In: Smartphones from an applied research perspective. InTech, Rijeka, Croatia, pp 1–24
6. LeMoyne R, Tomycz N, Mastroianni T, McCandless C, Cozza M, Peduto D (2015) Implementation of a smartphone wireless accelerometer platform for establishing deep brain stimulation treatment efficacy of essential tremor with machine learning. In: 37th annual international conference of the IEEE, Engineering in Medicine and Biology Society (EMBS), pp 6772–6775
7. LeMoyne R, Mastroianni T, McCandless C, Currivan C, Whiting D, Tomycz N (2018) Implementation of a smartphone as a wearable and wireless accelerometer and gyroscope platform for ascertaining deep brain stimulation treatment efficacy of Parkinson's disease through machine learning classification. Adv Park Dis 7(2):19–30
8. LeMoyne R, Mastroianni T, Tomycz N, Whiting D, Oh M, McCandless C, Currivan C, Peduto D (2017) Implementation of a multilayer perceptron neural network for classifying deep brain stimulation in 'On' and 'Off' modes through a smartphone representing a wearable and wireless sensor application. In: 47th Society for Neuroscience annual meeting (featured in Hot Topics; top 1% of abstracts)
9. LeMoyne R, Mastroianni T, Tomycz N, Whiting D, McCandless C, Peduto D, Cozza M (2015) I-Phone wireless accelerometer quantification of extremity tremor in essential tremor patient

undergoing activated and inactivated deep brain stimulation. In: International Neuromodulation Society's 12th World Congress

10. Hall M, Frank E, Holmes G, Pfahringer B, Reutemann P, Witten IH (2009) The WEKA data mining software: an update. ACM SIGKDD Explor Newsl 11(1):10–18

11. Witten IH, Frank E, Hall MA (2011) Data mining: practical machine learning tools and techniques. Morgan Kaufmann, Burlington, MA

12. WEKA [http://www.cs.waikato.ac.nz/~ml/weka/]

13. LeMoyne R, Kerr W, Mastroianni T, Hessel A (2014) Implementation of machine learning for classifying hemiplegic gait disparity through use of a force plate. In: 13th International Conference on Machine Learning and Applications (ICMLA), IEEE, pp 379–382

14. LeMoyne R, Kerr W, Zanjani K, Mastroianni T (2014) Implementation of an iPod wireless accelerometer application using machine learning to classify disparity of hemiplegic and healthy patellar tendon reflex pair. J Med Imaging Health Informatics 4(1):21–28

15. LeMoyne R, Mastroianni T, Hessel A, Nishikawa K (2015) Ankle rehabilitation system with feedback from a smartphone wireless gyroscope platform and machine learning classification. In: 14th International Conference on Machine Learning and Applications (ICMLA), IEEE, pp 406–409

16. LeMoyne R, Mastroianni T (2016) Implementation of a smartphone as a wireless gyroscope platform for quantifying reduced arm swing in hemiplegic gait with machine learning classification by multilayer perceptron neural network. In: 38th Annual international conference of the IEEE, Engineering in Medicine and Biology Society (EMBS), pp 2626–2630

17. LeMoyne R, Mastroianni T (2016) Smartphone wireless gyroscope platform for machine learning classification of hemiplegic patellar tendon reflex pair disparity through a multilayer perceptron neural network. In: Wireless Health (WH) of IEEE, pp 103–108

18. LeMoyne R, Mastroianni T (2017) Implementation of a smartphone wireless gyroscope platform with machine learning for classifying disparity of a hemiplegic patellar tendon reflex pair. J Mech Med Biol 17(6):1750083

19. LeMoyne R, Heerinckx F, Aranca T, De Jager R, Zesiewicz T, Saal HJ (2016) Wearable body and wireless inertial sensors for machine learning classification of gait for people with Friedreich's ataxia. In: IEEE 13th International conference on wearable and implantable Body Sensor Networks (BSN), pp 147–151

20. LeMoyne R, Mastroianni T (2015) Machine learning classification of a hemiplegic and healthy patellar tendon reflex pair through an iPod wireless gyroscope platform. In: 45th Society for Neuroscience annual meeting

21. Begg R, Kamruzzaman J (2005) A machine learning approach for automated recognition of movement patterns using basic, kinetic and kinematic gait data. J Biomech 38(3):401–408

22. Munakata T (2008) Fundamentals of the new artificial intelligence: neural, evolutionary, fuzzy and more. Springer, London

23. LeMoyne R, Mastroianni T (2016) Implementation of a multilayer perceptron neural network for classifying a hemiplegic and healthy reflex pair using an iPod wireless gyroscope platform. In: 46th Society for Neuroscience annual meeting

24. LeMoyne R, Mastroianni T, Hessel A, Nishikawa K (2015) Application of a multilayer perceptron neural network for classifying software platforms of a powered prosthesis through a force plate. In: 14th International Conference on Machine Learning and Applications (ICMLA), IEEE, pp 402–405

25. LeMoyne R, Mastroianni T (2017) Virtual proprioception for eccentric training. In: 39th annual international conference of the IEEE, Engineering in Medicine and Biology Society (EMBS), pp 4557–4561

26. LeMoyne R, Mastroianni T (2017) Wireless gyroscope platform enabled by a portable media device for quantifying wobble board therapy. In: 39th Annual international conference of the IEEE, Engineering in Medicine and Biology Society (EMBS), pp 2662–2666

27. LeMoyne R, Mastroianni T, McCandless C, Currivan C, Whiting D, Tomycz N (2018) Implementation of a smartphone as a wearable and wireless inertial sensor platform for determining efficacy of deep brain stimulation for Parkinson's disease tremor through machine learning. In: 48th Society for Neuroscience annual meeting (Nanosymposium)

Chapter 9
Assessment of Machine Learning Classification Strategies for the Differentiation of Deep Brain Stimulation "On" and "Off" Status for Parkinson's Disease Using a Smartphone as a Wearable and Wireless Inertial Sensor for Quantified Feedback

Abstract The considerable advantage of integrating wearable and wireless systems with machine learning for the assessment of deep brain stimulation parameter configuration status is addressed. A wearable and wireless system, such as a smartphone with its accelerometer and gyroscope, provides the quantified basis for the efficacy determination of a treatment strategy. In particular, deep brain stimulation is well suited for being amalgamated with wearable and wireless systems. For a subject with Parkinson's disease, deep brain stimulation set to "On" and "Off" status is distinguished through an assortment of machine learning algorithms, such as J48 decision tree, K-nearest neighbors, logistic regression, support vector machine, multilayer perceptron neural network, and random forest. The feature set is consolidated from the accelerometer signal and gyroscope signal from a smartphone using software automation. The appropriateness for these machine learning algorithms was assessed in terms of both classification accuracy and computational efficiency. These capabilities further refine the opportunities of machine learning classification being allocated local to the wearable and wireless system with an available Cloud computing resource. These findings establish a preliminary perspective regarding the utility of Network Centric Therapy, for which effectively real-time optimization of parameter configurations for deep brain stimulation can be developed in a patient-specific context. Furthermore, the real-time optimization process can be adaptive to the inherent temporal fluctuations of a progressive neurodegenerative movement disorder, such as Parkinson's disease and Essential tremor.

Keywords Deep brain stimulation · Parameter configuration optimization · Wearable and wireless systems · Quantified feedback · Machine learning · Classification accuracy · J48 decision tree · K-nearest neighbors · Logistic regression · Support vector machine · Multilayer perceptron neural network · Random forest · Network Centric Therapy

© Springer Nature Singapore Pte Ltd. 2019   113
R. LeMoyne et al., *Wearable and Wireless Systems for Healthcare II*,
Smart Sensors, Measurement and Instrumentation 31,
https://doi.org/10.1007/978-981-13-5808-1_9

9.1 Introduction

Multiple research, development, testing, and evaluation endeavors have progressively demonstrated the utility and feasibility of amalgamating wearable and wireless systems, machine learning, and deep brain stimulation for the treatment of neurodegenerative movement disorders, such as Parkinson's disease and Essential tremor. The wearable and wireless system, such as a smartphone equipped with an accelerometer and gyroscope, provides objectively quantified feedback for response to deep brain stimulation. With machine learning the ability to attain classification accuracy between two configurations for the deep brain stimulation system, such as "On" and "Off," is enabled [1–3].

   Machine learning classification, such as through the Waikato Environment for Knowledge Analysis (WEKA), requires wearable and wireless inertial sensor system signal data to be consolidated into a feature set. The consolidation of the feature set from the inertial signal data is achieved through the application of a software automation program. This aspect of the post-processing phase can be readily applied in a remote setting relative to the location of the acquisition of the experimental data [1–3].

   Preliminary Network Centric Therapy is demonstrated through the success of these proof-of-concept endeavors as illustrated in Fig. 9.1. The wearable and wireless system for quantifying response to deep brain stimulation setting, such as "On" of "Off," is provided by a smartphone. The smartphone can convey the recorded inertial sensor signal data through wireless connectivity to the Internet as an email attachment. The email resource constitutes the functional semblance of a Cloud computing environment. From the email resource, the inertial sensor signal data acquired by the smartphone can be downloaded anywhere in the world for post-processing. An automated software routine consolidates the signal data to an appropriate feature set, for which machine learning attains a classification accuracy to differentiate between two scenarios, such as deep brain stimulation set to "On" and "Off" mode [1–3]. Wearable and wireless systems in conjunction with machine

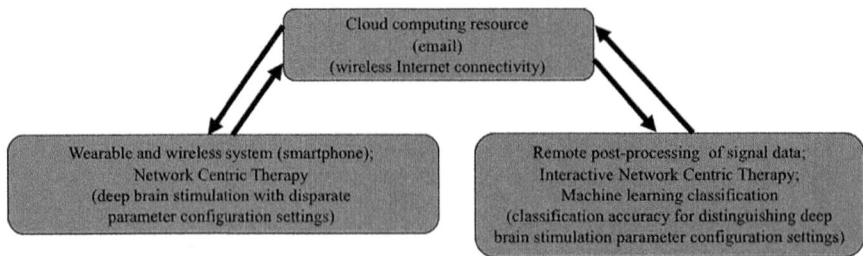

Fig. 9.1 Schematic representing the process for preliminary Network Centric Therapy for the treatment of neurodegenerative movement disorders, such as Parkinson's disease and Essential tremor, through deep brain stimulation, a wearable and wireless system, such as a smartphone, and machine learning

learning have been proposed to enable the ability to ascertain optimal parameter configurations for deep brain stimulation tuning [1–8].

However, there are many machine learning algorithms available for the determination of optimal deep brain stimulation tuning. Furthermore, machine learning algorithms are contextually appropriate based on the characteristics of the scenario being classified [9]. Six machine learning algorithms are of particular interest:

- J48 decision tree
- K-nearest neighbors
- Logistic regression
- Support vector machine
- Multilayer perceptron neural network
- Random forest

These machine learning algorithms are readily provided by WEKA [10–12].

The research objective elucidates an assortment of machine learning algorithms for the classification accuracy of deep brain stimulation set to "On" and "Off" status using a smartphone as a wearable and wireless inertial sensor system. Also elucidated is the amenability for being integrated into the optimization process regarding the parameter configuration of the deep brain stimulation system. The goal is envisioned to facilitate the progressive advance to achieve closed-loop optimization of deep brain stimulation system parameter configuration tuning.

9.2 Background

The concept of deep brain stimulation has been reviewed in Chap. 4 "Deep brain stimulation for the treatment of movement disorder regarding Parkinson's disease and Essential tremor device characterization." The efficacy of deep brain stimulation for treating neurodegenerative movement disorders, such as Parkinson's disease, extends beyond the domain of an expert neurosurgeon [3, 13]. The tuning of a deep brain stimulation system requires consideration of multiple parameters, such as amplitude, electrode polarity, frequency, and pulse width [14, 15]. The acquisition of an optimal deep brain stimulation parameter configuration is imperative [13].

However, the optimal programming of the deep brain stimulation parameters can present a time-consuming task. A considerable number of permutations exist regarding the parameter configurations. An organized strategy for converging to an optimal parameter configuration would advance the treatment of people with deep brain stimulation to treat neurodegenerative movement disorders [13].

Such an advanced concept for achieving optimal deep brain stimulation parameter configurations that are specific to the patient would considerably improve healthcare for people with neurodegenerative movement disorders. Strain on critical and limited medical resources could be alleviated, especially if the technique could be automated. Objectively quantified feedback to measure and record tremor status would serve as a critical role for the automation process. Inertial sensors intrinsic to wearable and wireless systems, such as a smartphone, enable the ability to robustly quantify movement disorder tremor [4–8].

During 2010 LeMoyne et al. demonstrated the feasibility of applying the smartphone as a wireless accelerometer platform that is essentially wearable for the quantification of Parkinson's disease hand tremor. The accelerometer signal data would be conveyed to from the experimental site to a post-processing resource by wireless transmission to the Internet as an email attachment [16]. This capability is a central theme to the concept of Network Centric Therapy as experimental and post-processing resources can be remotely positioned anywhere in the world, respective of each other.

Further progressive evolution was achieved when machine learning was applied to classify states of deep brain stimulation, such as "On" and "Off." The feature set was derived from consolidating the inertial sensor signal data from a smartphone functioning as a wearable and wireless system. Using machine learning algorithms provided by WEKA, considerable classification accuracy to distinguish between deep brain stimulation "On" and "Off" settings for both Parkinson's disease and Essential tremor was achieved [1–3].

WEKA consists of many machine learning algorithms, such as J48 decision tree, K-nearest neighbors, logistic regression, support vector machine, multilayer perceptron neural network, and random forest. Each machine learning algorithm comprises a unique strategy for achieving its respective classification accuracy. Furthermore, the computational complexity is unique to the specific machine learning algorithm under consideration [10–12]. These characteristics of machine learning algorithms underscore the utility of contrasting multiple machine learning algorithms. The future objective is to apply the most appropriate machine learning technique for Network Centric Therapy to facilitate the ability to ascertain patient-specific optimal parameter configurations for deep brain stimulation treatment of neurodegenerative movement disorders.

9.3   Method and Materials

The data incorporated to achieve the research objective is derived from "Implementation of a smartphone as a wearable and wireless accelerometer and gyroscope platform for ascertaining deep brain stimulation treatment efficacy of Parkinson's disease through machine learning classification" published in *Advances in Parkinson's Disease* with further machine learning classification analysis applied to the 48th Society for Neuroscience Nanosymposium titled "Implementation of a smartphone as a wearable and wireless inertial sensor platform for determining efficacy of deep brain stimulation for Parkinson's disease tremor through machine learning." An engineering proof-of-concept perspective is provided with respect to applying machine learning to differentiate between "On" and "Off" status regarding deep brain stimulation for ameliorating Parkinson's disease tremor symptoms. The one subject for the experiment was treated with bilateral subthalamic nucleus deep brain stimulation. Informed consent was established, and the experiment was conducted at Allegheny General Hospital [3, 17].

The wearable and wireless system for quantifying the Parkinson's disease tremor was achieved by mounting a smartphone to the dorsum of the hand through a latex glove as illustrated in Fig. 9.2. The accelerometer and gyroscope signal data acquired by the smartphone was wirelessly transmitted to the Internet as an email attachment. The email resource constituted a functional Cloud computing environment that enabled remote post-processing anywhere in the world [3, 17].

For the post-processing phase, the implementation of a robust software automation technique substantially reduced strain on resources. Furthermore, the inherent nature of software automation facilitates the consistency of reducing data from the inertial sensor signals, such as from the accelerometer and gyroscope. From a global perspective, Matlab shall be applied to post-process the inertial sensor signal data to an Attribute-Relation File Format (ARFF) file [3, 17]. Python has also been successfully demonstrated to consolidate inertial sensor signal data for machine learning in an automated manner [18]. The ARFF file shall then be uploaded to the Waikato Environment for Knowledge Analysis (WEKA) for multiple machine learning algorithms to ascertain their classification accuracy. Figure 9.3 summarizes the global post-processing perspective [3, 17].

The development of the software automation program is first commenced through the use of pseudo code. The pseudo code provides a preliminary insight of the primary functional requirements of the software program before actual coding [19]. As summarized in Fig. 9.4, the pseudo code for the automated post-processing of the inertial sensor signal data is read into a vectorized format, vectorized signal data is consolidated into descriptive statistics for the ARFF file feature set, and then the numeric attributes are applied to the formatting parameters required for the ARFF file. Further assurance of the software program can be established through the Fagan inspection process [20].

Fig. 9.2 Smartphone mounted to the dorsum of the hand for a subject [3, 17]

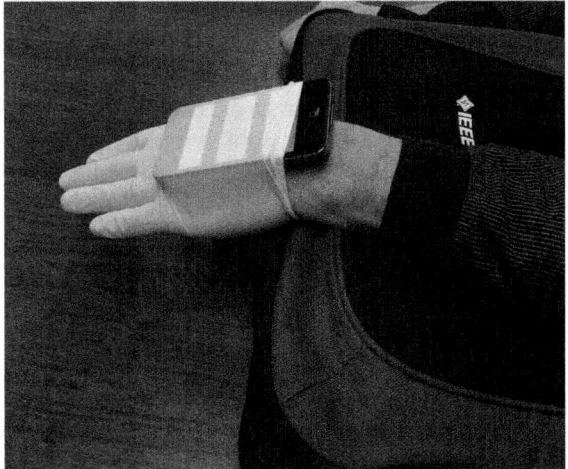

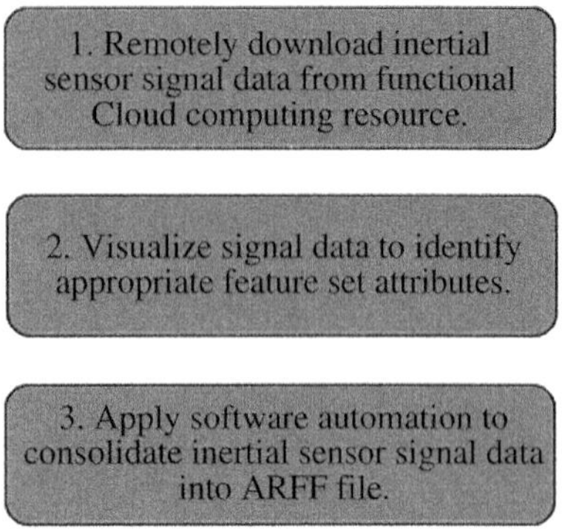

Fig. 9.3 Global post-processing perspective for reducing the inertial sensor signal data through software automation to an ARFF file for machine learning classification with WEKA [3, 17]

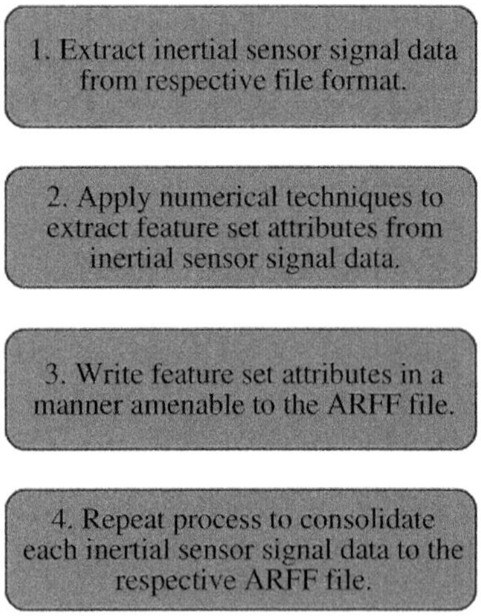

1. Remotely download inertial sensor signal data from functional Cloud computing resource.

2. Visualize signal data to identify appropriate feature set attributes.

3. Apply software automation to consolidate inertial sensor signal data into ARFF file.

Fig. 9.4 Pseudo code for automated post-processing of the inertial sensor signal data to an ARFF file [3, 17]

1. Extract inertial sensor signal data from respective file format.

2. Apply numerical techniques to extract feature set attributes from inertial sensor signal data.

3. Write feature set attributes in a manner amenable to the ARFF file.

4. Repeat process to consolidate each inertial sensor signal data to the respective ARFF file.

The experimental protocol applied ten repetitions involving the deep brain stimulation system set to "On" status and ten repetitions involving the deep brain stimulation system set to "Off" status. The experimental trials were accomplished with the subject's wrist suspended beyond an arm support. The smartphone was equipped with an application that simultaneously recorded the accelerometer and gyroscope signal at a sampling rate of 100 Hz for an appropriate duration of 10 s. The following process in Fig. 9.5 illustrates the experimental protocol [3, 17].

9.4   Results and Discussion

For the subject with Parkinson's disease, the deep brain stimulation system perceptibly suppressed tremor regarding the "On" setting relative to the "Off" setting. The observational perception is transcended with the objectively quantified inertial sensor signal from the accelerometer and gyroscope provided by the smartphone as a wearable and wireless system. Figure 9.6 represents the acceleration magnitude for

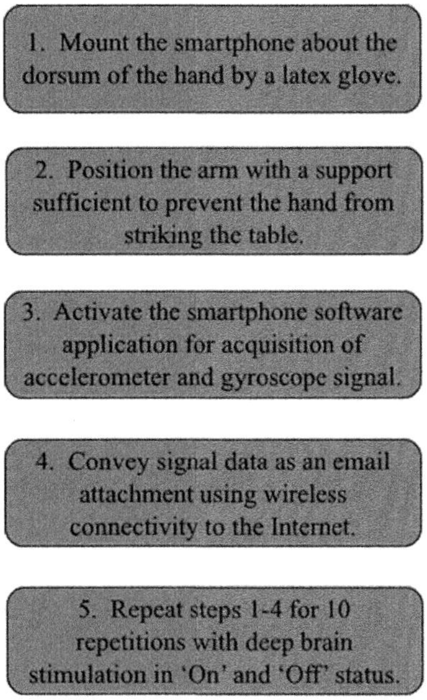

Fig. 9.5 Experimental protocol for acquisition of inertial sensor signal data (accelerometer and gyroscope) through a smartphone as a wearable and wireless system with deep brain stimulation system set to "On" and "Off" status [3, 17]

1. Mount the smartphone about the dorsum of the hand by a latex glove.

2. Position the arm with a support sufficient to prevent the hand from striking the table.

3. Activate the smartphone software application for acquisition of accelerometer and gyroscope signal.

4. Convey signal data as an email attachment using wireless connectivity to the Internet.

5. Repeat steps 1-4 for 10 repetitions with deep brain stimulation in 'On' and 'Off' status.

the subject with the deep brain stimulation system in "On" mode. Figure 9.7 presents the acceleration magnitude for the subject with the deep brain stimulation system in "Off" mode. Figures 9.6 and 9.7 are also visibly disparate. Therefore, machine learning classification is a plausible technique for distinguishing between deep brain stimulation "On" and "Off" mode [3, 17].

The feature set comprised four signals based on the inertial sensors of the smartphone, such as acceleration magnitude and the roll, pitch, and yaw aspects derived from the gyroscope signal. Five descriptive statistic attributes were acquitted for each of the four inertial sensor signals:

- Maximum
- Minimum
- Mean
- Standard deviation
- Coefficient of variation [3, 17]

Using the previously described software automation technique, the inertial sensor signal data was consolidated into a feature set. The following machine learning algorithms were evaluated for classification accuracy and suitability for being integrated into the parameter configuration optimization process of the deep brain stimulation system:

- J48 decision tree
- K-nearest neighbors
- Logistic regression

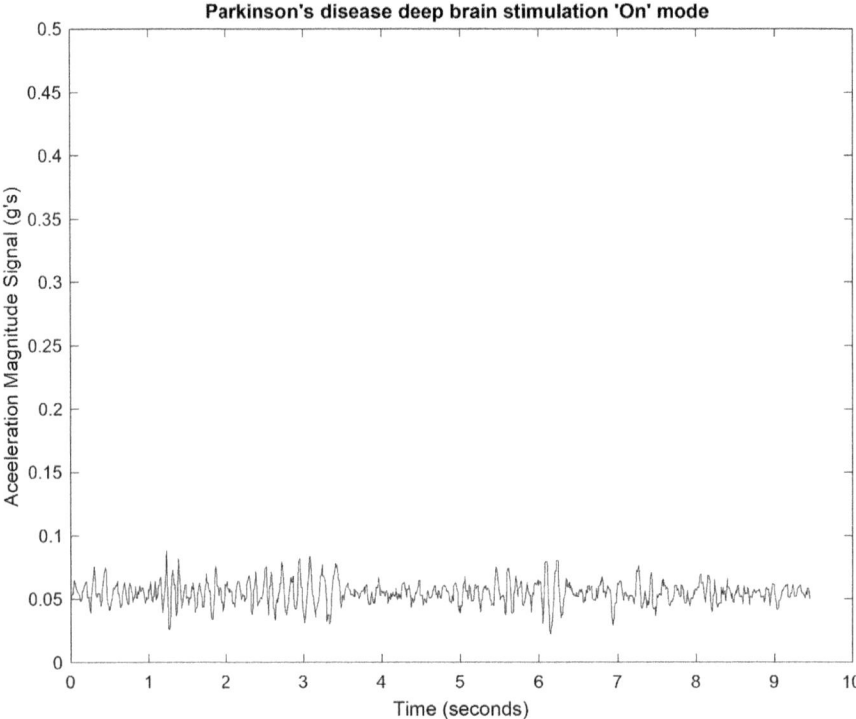

Fig. 9.6 Acceleration magnitude signal quantified through a smartphone as a wearable and wireless inertial sensor system for a subject with Parkinson's disease and deep brain stimulation system set to "On" mode [3, 17]

- Support vector machine
- Multilayer perceptron neural network
- Random forest [3, 17]

Figure 9.8 summarizes the classification accuracy of these machine learning algorithms for their ability to distinguish between deep brain stimulation "On" and "Off" status for a subject with Parkinson's disease based on the quantified inertial sensor signals provided by a smartphone functioning as a wearable and wireless system. The application of these six machine learning algorithms ascertains that logistic regression, support vector machine, and multilayer perceptron neural network attain 95% classification accuracy, which is the greatest [3, 17]. Further investigation of the machine learning classification endeavor can be elucidated by considering the confusion matrix.

The confusion matrix is presented as the number of correctly classified instances contrasted to the number of incorrectly classified instances. With respect to binary classification, the confusion matrix is represented by a two-by-two matrix [10–12]. Logistic regression, support vector machine, and multilayer perceptron neural network achieved the greatest classification accuracy. For support vector machine and

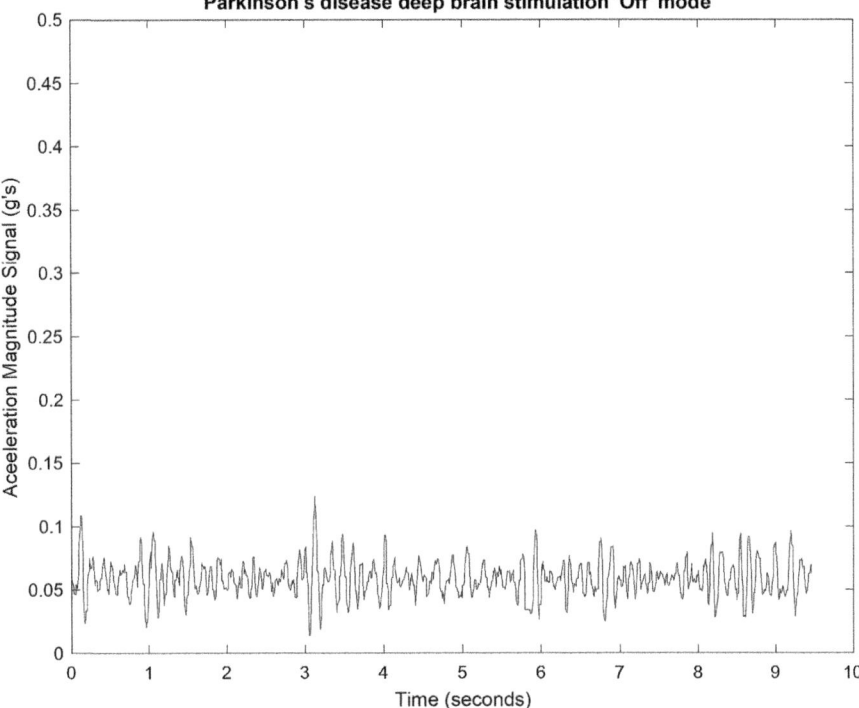

Fig. 9.7 Acceleration magnitude signal quantified through a smartphone as a wearable and wireless inertial sensor system for a subject with Parkinson's disease and deep brain stimulation system set to "Off" mode [3, 17]

multilayer perceptron neural network, one deep brain stimulation set to "On" instance was incorrectly classified as deep brain stimulation set to "Off." Regarding logistic regression one deep brain stimulation trial set to "Off" instance was misclassified as being deep brain stimulation set to "On."

Another consideration for the suitability of applying a machine learning algorithm is based on the time to determine the classification accuracy. Figure 9.9 presents the time elapse for processing the support vector machine, logistic regression, and multilayer perceptron neural network. Notably, the support vector machine achieves 95% classification accuracy in 0.02 s, which is 12 times faster the multilayer perceptron neural network. Furthermore, logistic regression attains 95% classification accuracy in only 0.03 s, which is eight times faster the multilayer perceptron neural network. The significance of the processing time duration requirement is highly dependent of the wireless architecture desired.

Notably the processing time is derived from the operation of a conventional personal computer. A local microprocessor integrated into a wearable and wireless system may be relatively more constrained for conducting machine learning classification. In this scenario the application of the support vector machine would be

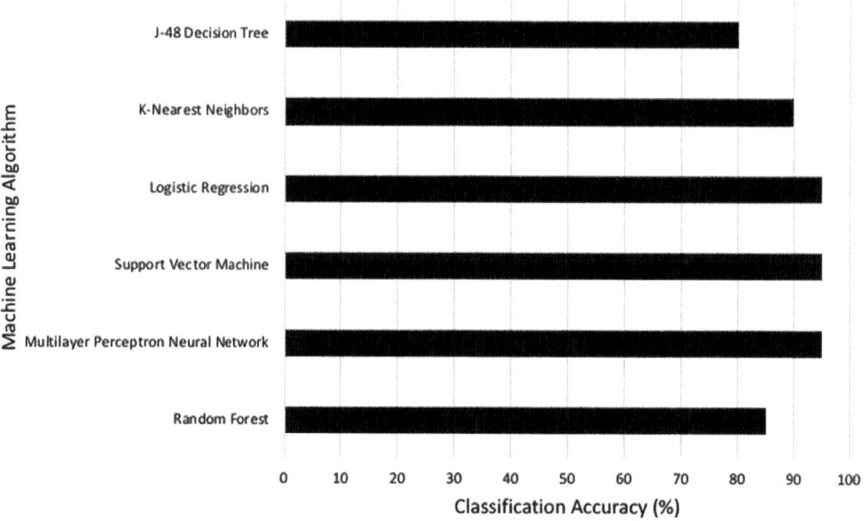

Fig. 9.8 Machine learning classification accuracy for J48 decision tree, K-nearest neighbors, logistic regression, support vector machine, multilayer perceptron neural network, and random forest to differentiate between deep brain stimulation "On" and "Off" status for a subject with Parkinson's disease using a smartphone as a wearable and wireless inertial sensor system [3, 17]

beneficial. However, if the machine learning classification was post-processed by a Cloud computing resource, with a computational processing capability transcending the respective personal computer, concern for the machine learning classification algorithm processing time could be alleviated.

9.5   Network Centric Therapy Integrating Wearable and Wireless Systems as Quantified Feedback for Deep Brain Stimulation Using Machine Learning Classification

There are two available conceptual architectures for the amalgamation of deep brain stimulation parameter configuration optimization established by wearable and wireless systems for quantified feedback using machine learning classification. The appropriateness of these two conceptual architectures is highly dependent on the processing time requirements imparted by the machine learning algorithm. Figures 9.10 and 9.11 visualize conceptual architectures applying machine learning classification from the wearable and wireless system and Cloud computing resource with associated remote post-processing.

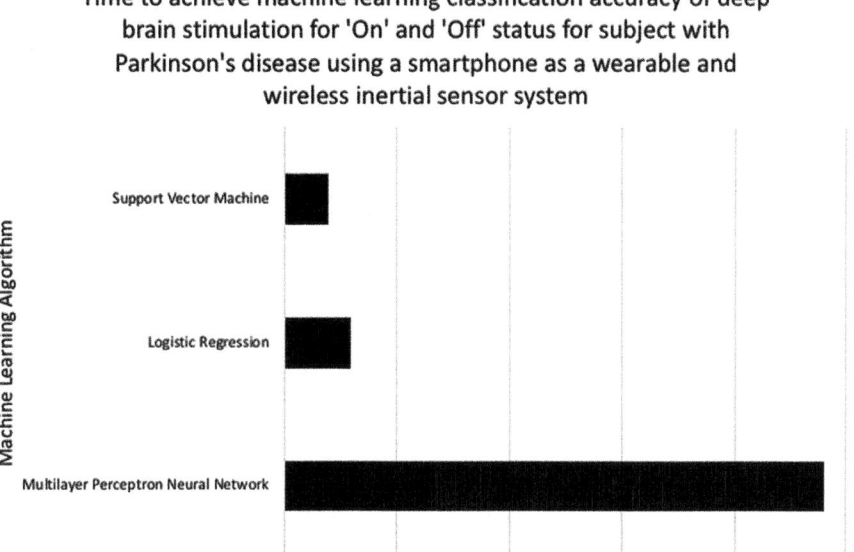

Fig. 9.9 Machine learning classification accuracy processing time elapse for support vector machine, logistic regression, and multilayer perceptron neural network [3, 17]

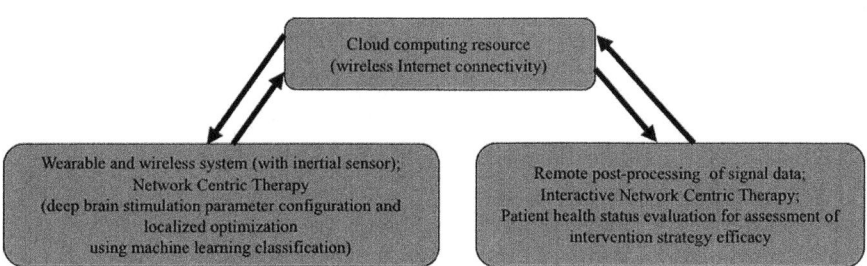

Fig. 9.10 Optimization of deep brain stimulation parameter configuration achieved through machine learning classification by the wearable and wireless system with patient health status later conveyed to Cloud computing resource

These research findings establish a prelude to the utility of Network Centric Therapy. Wearable and wireless systems provide the quantified foundation respective of the efficacy of a therapy intervention for the treatment of a neurodegenerative disorder, such as Parkinson's disease and Essential tremor. This quantified feedback can be intrinsically applied to ascertaining an optimal parameter configuration for deep brain stimulation with machine learning serving as a basis for converging upon the optimal status. Depending on the desired response time for real-time acquisition of an optimal parameter configuration, the machine learning platform can be locally

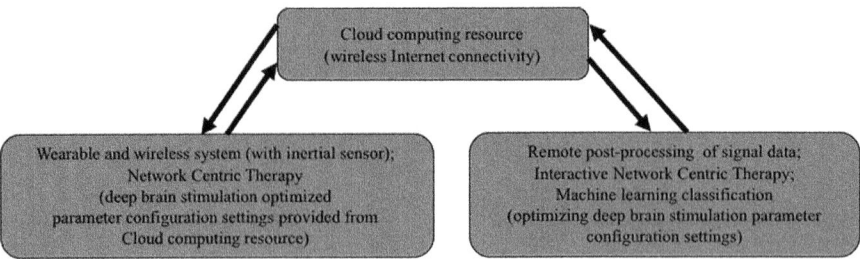

Fig. 9.11 Optimization of deep brain stimulation parameter configuration achieved through machine learning classification determined by Cloud computing resource based on wearable and wireless system quantified feedback conveyed to Cloud computing resource with associated remote post-processing

situated with the wearable and wireless system or reside with the Cloud computing resources. The implications are this form of Network Centric Therapy can facilitate patient-specific optimization of a deep brain stimulation parameter configuration while also providing the capacity to adjust to the inherent real-time variations of neurodegenerative movement disorders, such as Parkinson's disease and Essential tremor.

9.6   Conclusion

The successful amalgamation of a wearable and wireless system, machine learning, and deep brain stimulation has been presented. The wearable and wireless system is enabled by a smartphone. The inertial sensor internal to the smartphone consists of both an accelerometer and gyroscope, for which their signal data is conveyed as an email attachment. The email resource is considered representative of a functional Cloud computing resource.

The wearable and wireless system provided by a smartphone quantified two perceptibly distinguishable deep brain stimulation statuses, such as "On" and "Off," for a subject with Parkinson's disease. The signal data is processed by software automation to an ARFF file for machine learning classification by WEKA. The feature set is comprised of the acceleration magnitude signal and three orthogonal signals acquired from the gyroscope. Each of these four signals are each consolidated into five attributes, such as maximum, minimum, mean, standard deviation, and coefficient of variation.

Six machine learning algorithms are evaluated: J48 decision tree, K-nearest neighbors, logistic regression, support vector machine, multilayer perceptron neural network, and random forest. These machine learning algorithms attain considerable classification accuracy. However, in terms of both classification accuracy and computational efficiency, support vector machine achieves the best performance. These findings constitute a prelude to the capabilities of Network Centric Therapy, such as the real-time acquisition of optimal parameter configurations for deep brain stimulation that are both patient-specific and adaptive to the inherently real-time fluctuations of neurodegenerative movement disorders, such as Parkinson's disease and Essential tremor.

References

1. LeMoyne R, Tomycz N, Mastroianni T, McCandless C, Cozza M, Peduto D (2015) Implementation of a smartphone wireless accelerometer platform for establishing deep brain stimulation treatment efficacy of essential tremor with machine learning. In: 37th annual international conference of the IEEE, Engineering in Medicine and Biology Society (EMBS), pp 6772–6775
2. LeMoyne R, Mastroianni T, Tomycz N, Whiting D, Oh M, McCandless C, Currivan C, Peduto D (2017) Implementation of a multilayer perceptron neural network for classifying deep brain stimulation in 'On' and 'Off' modes through a smartphone representing a wearable and wireless sensor application. In: 47th Society for Neuroscience annual meeting (featured in Hot Topics; top 1% of abstracts)
3. LeMoyne R, Mastroianni T, McCandless C, Currivan C, Whiting D, Tomycz N (2018) Implementation of a smartphone as a wearable and wireless accelerometer and gyroscope platform for ascertaining deep brain stimulation treatment efficacy of Parkinson's disease through machine learning classification. Adv Park Dis 7(2):19–30
4. LeMoyne R, Coroian C, Cozza M, Opalinski P, Mastroianni T, Grundfest W (2009) The merits of artificial proprioception, with applications in biofeedback gait rehabilitation concepts and movement disorder characterization. In: Biomedical engineering. InTech, Vienna, pp 165–198
5. LeMoyne R, Mastroianni T (2017) Smartphone and portable media device: a novel pathway toward the diagnostic characterization of human movement. In: Smartphones from an applied research perspective. InTech, Rijeka, Croatia, pp 1–24
6. LeMoyne R, Mastroianni T (2017) Wearable and wireless gait analysis platforms: smartphones and portable media devices. In: Wireless MEMS networks and applications. Elsevier, New York, pp 129–152
7. LeMoyne R, Mastroianni T (2016) Telemedicine perspectives for wearable and wireless applications serving the domain of neurorehabilitation and movement disorder treatment. In: Telemedicine, SMGroup, Dover, Delaware, pp 1–10
8. LeMoyne R, Mastroianni T (2015) Use of smartphones and portable media devices for quantifying human movement characteristics of gait, tendon reflex response, and Parkinson's disease hand tremor. In: Mobile health technologies, methods and protocols. Springer, New York, pp 335–358
9. LeMoyne R, Kerr W, Mastroianni T, Hessel A (2014) Implementation of machine learning for classifying hemiplegic gait disparity through use of a force plate. In: 13th International Conference on Machine Learning and Applications (ICMLA), IEEE, pp 379–382
10. Hall M, Frank E, Holmes G, Pfahringer B, Reutemann P, Witten IH (2009) The WEKA data mining software: an update. ACM SIGKDD Explor Newsl 11(1):10–18
11. Witten IH, Frank E, Hall MA (2011) Data mining: practical machine learning tools and techniques. Morgan Kaufmann, Burlington, MA
12. WEKA [http://www.cs.waikato.ac.nz/~ml/weka/]
13. Isaias IU, Tagliati M (2008) Deep brain stimulation programming for movement disorders. In: Deep brain stimulation in neurological and psychiatric disorders. Springer, New York, pp 361–397
14. Volkmann J, Moro E, Pahwa R (2006) Basic algorithms for the programming of deep brain stimulation in Parkinson's disease. Mov Disord 21(S14):S284–S289

15. Amon A, Alesch F (2017) Systems for deep brain stimulation: review of technical features. J Neural Transm 124(9):1083–1091
16. LeMoyne R, Mastroianni T, Cozza M, Coroian C, Grundfest W (2010) Implementation of an iPhone for characterizing Parkinson's disease tremor through a wireless accelerometer application. In: 32nd annual international conference of the IEEE, Engineering in Medicine and Biology Society (EMBS), pp 4954–4958
17. LeMoyne R, Mastroianni T, McCandless C, Currivan C, Whiting D, Tomycz N (2018) Implementation of a smartphone as a wearable and wireless inertial sensor platform for determining efficacy of deep brain stimulation for Parkinson's disease tremor through machine learning. In: 48th Society for Neuroscience annual meeting (Nanosymposium)
18. LeMoyne R, Heerinckx F, Aranca T, De Jager R, Zesiewicz T, Saal HJ (2016) Wearable body and wireless inertial sensors for machine learning classification of gait for people with Friedreich's ataxia. In: IEEE 13th international conference on wearable and implantable Body Sensor Networks (BSN), pp 147–151
19. Sommerville I (2011) Software engineering. Addison-Wesley, Boston
20. Fagan ME (1999) Design and code inspections to reduce errors in program development. IBM Syst J 38(2/3):258–287

Chapter 10
New Perspectives for Network Centric Therapy for the Treatment of Parkinson's Disease and Essential Tremor

Abstract The evolutionary trends of wearable and wireless systems for the treatment of progressive neurodegenerative movement disorders, such as Parkinson's disease and Essential tremor, have been elucidated. Wearable and wireless systems further synergize with advanced technologies, such as deep brain stimulation. These capabilities establish the basis for Network Centric Therapy for progressive neurodegenerative movement disorders, such as Parkinson's disease and Essential tremor.

Keywords Network Centric Therapy · Wearable and wireless systems · Parkinson's disease · Essential tremor · Deep brain stimulation · Cloud computing · Data science · Optimization · Parameter configuration

Network Centric Therapy is a representative perspective of the Internet of Things for healthcare. In particular the scope of this book pertains to the treatment of two prevalent types of movement disorders, such as Parkinson's disease and Essential tremor. Wearable and wireless systems, such as a smartphone, constitute a sensor level representation of the Internet of Things for this domain of healthcare. As opposed to traditional and subjective ordinal scale methodologies to diagnosing the efficacy of a treatment strategy, wearable and wireless inertial sensor systems can quantify the response to a prescribed intervention. In particular, the application of these wearable and wireless systems does not require the application of highly specialized clinical resources. The response to therapy can be provided at the convenience of the subject's homebound setting, and the trial data can be provided to clinical post-processing resources through wireless connectivity to the Internet.

Deep brain stimulation offers and advanced technique for the treatment of progressive neurodegenerative movement disorders, such as Parkinson's disease and Essential tremor. An issue to the application of deep brain stimulation is the proper acquisition of an effective parameter configuration. The convergence of a deep brain stimulation system parameter configuration can be challenging, especially with respect to the quantity of available permutations.

The amalgamation of deep brain stimulation system with the wearable and wireless inertial sensor system facilitates the parameter configuration convergence with

© Springer Nature Singapore Pte Ltd. 2019   127
R. LeMoyne et al., *Wearable and Wireless Systems for Healthcare II*,
Smart Sensors, Measurement and Instrumentation 31,
https://doi.org/10.1007/978-981-13-5808-1_10

objectively quantified signal data. The inertial sensor signal data can be further consolidated into a feature set that can utilize machine learning for the acquisition of a real-time optimized deep brain stimulation parameter configuration, such as from a Cloud computing resource provided by Network Centric Therapy. The implication is that a subject with a progressive neurodegenerative movement disorders, such as Parkinson's disease and Essential tremor, can receive highly interactive therapy intervention from the convenience of a homebound and autonomous environment for highly skilled clinical experts anywhere in the world.

As a general example, a subject with Parkinson's disease in remote Alaska could receive high-quality treatment for world-renowned clinicians from Pittsburgh, Pennsylvania. An automated process made available by Network Centric Therapy could optimize in a real-time context the parameter configuration for a deep brain stimulation system. At the discretion of the expert clinical team in Pittsburgh, the parameter configuration bounds could be modified for a more appropriate treatment strategy. Network Centric Therapy would enable highly interactive therapy between the subject and clinician, and the robust integration of data science can further advance treatment strategies.